Quantitative Nuclear Cardiography

Quantitative Nuclear Cardiography

Edited by

RICHARD N. PIERSON, JR., M.D.
Columbia University College of Physicians and Surgeons
St. Luke's Hospital Center, New York,
and Hackensack Hospital,
Hackensack, New Jersey

JOSEPH P. KRISS, M.D.
Stanford University Medical Center
Palo Alto, California

ROBERT H. JONES, M.D.
Duke University Medical Center
Durham, North Carolina

WILLIAM J. MACINTYRE, PH.D.
Cleveland Clinic
Cleveland, Ohio

A WILEY BIOMEDICAL-HEALTH PUBLICATION

John Wiley & Sons, *New York • London • Sydney • Toronto*

THIS TEXT IS DEDICATED TO OUR WIVES,
OUR STAFFS, AND OUR PATIENTS

Copyright © 1975, by John Wiley & Sons, Inc.

All rights reserved. Published simultaneously in Canada.

No part of this book may be reproduced by any means, nor transmitted, nor translated into a machine language without the written permission of the publisher.

Library of Congress Cataloging in Publication Data

Main entry under title:

Quantitative nuclear cardiography.

  (A Wiley biomedical-health publication)
  Includes bibliographical references and index.
  1. Radioisotopes in cardiology. I. Pierson, Richard N., ed. [DNLM: 1. Angiocardiography. 2. Heart—Radiography. WG141 Q15]

RC683.5.R3Q3   616.1'2'07575   74-20990
ISBN 0-471-68950-5

Printed in the United States of America

10 9 8 7 6 5 4 3 2 1

Authors

Anderson, Page A. W., Md., Duke University Medical Center, Durham, North Carolina

Ashburn, William, M.D., University of California at San Diego, San Diego, California

Bassingthwaighte, James B., M.D., Ph.D., Department of Physiology, Mayo Clinic and Foundation, Rochester, Minnesota

Bonte, Frederick J., M.D., The University of Texas Health Center at Dallas, Dallas, Texas

Budinger, Thomas, M.D., Ph.D., Donner Laboratory, University of California, Berkeley, California

Cannon, Paul, J., M.D., Columbia University College of Physicians and Surgeons, New York, New York

Castellana, Frank C., SCD (Eng.), Department of Chemical Engineering, Columbia University, New York, New York

Chervu, Rao, Ph.D., Albert Einstein Medical Center, New York, New York

Jones, Robert H., M.D., Duke University Medical Center, Durham, North Carolina

Kriss, Joseph, P., M.D., Stanford University Medical Center, Palo Alto, California

Lidofsky, Leon, Ph.D., Columbia University, New York, New York

MacIntyre, William J., Ph.D., Cleveland Clinic, Cleveland, Ohio

Parkey, Robert W., M.D., The University of Texas Health Center at Dallas, Dallas, Texas

Pierson, Richard N., Jr., M.D., Columbia University College of Physicians and Surgeons, St. Luke's Hospital Center, New York, and Hackensack Hospital, Hackensack, New Jersey

Van Dyke, Donald C., M.D., Donner Laboratory, University of California, Berkeley, California

Weber, Paul, M.D., Kaiser Permanente Hospital, Oakland, California

Acknowledgments

Figures 1 to 4 are reproduced by permission from *Seminars in Nuclear Medicine* **3**: 177, 1973.

Figures 5, 15, 16, and 17 are reproduced by permission from Measurement of Heart Chamber Volumes by Analysis of Dilution Curves Simultaneously Recorded by Scintillation Camera, *Circulation* **44**: 37–46, 1971, with approval of the American Heart Association.

Figures 6 to 14 are reproduced by permission from Dynamic Studies with Radioisotopes in Clinical Medicine and Research, *LAEA-SM*-185, Knoxville, 1972.

Figures 18 to 26 are reproduced by permission from the *American Journal of Cardiology* **16**: 165–175, 1965.

Figures 32 to 34 are reproduced from Quantitation of Left to Right Cardiac Shunts with Radionuclide Angiography, *Circulation* **49**: 512–516, 1974, with the approval of the American Heart Association.

Figures 27a and b, 28, 37, and 90 are reproduced from Radioisotopic Angiocardiography: Wide Scope of Applicability in Diagnosis and Evaluation and Therapy in Diseases of the Heart and Great Vessels, *Circulation* **43**: 792–808, 1871, with the approval of the American Heart Association.

Figures 29 to 31 and 38 to 43 are reproduced with permission from the *Journal of Nuclear Medicine* **13**: 31, 1972.

Figures 44 to 48 are reproduced with permission from the *Radiology Clinics of North America* **9**: 369, 1971.

Figure 49 is reproduced by permission from the *Journal of Nuclear Medicine* **11**: 723, 1970.

Figure 72 is reproduced with permission from the *American Journal of Medicine* **53**: 775, 1972.

Figures 99 to 111 are reproduced by permission from the *British Heart Journal* **36**: 122–131, 1974.

Figures 121 to 134 are reprinted by permission from the *Journal of Clinical Investigation* **51**: 964–994, 1972.

Contents

1 Introduction   1
Richard N. Pierson, Jr.

2 The Normal Heart   4
William J. MacIntyre, with Joseph P. Kriss

3 Congenital Heart Disease: Imaging and Analytic Methods   32
Robert H. Jones, with Page A. W. Anderson

4 Acquired Cardiovascular Disease   66
Joseph P. Kriss

5 Analysis of Left Ventricular Function   123
Richard N. Pierson Jr., with Donald C. Van Dyke

6 Measurements of Regional Myocardial Perfusion   155
William J. MacIntyre, with Paul J. Cannon and William Ashburn

7 Mathematical Modeling of the Central Circulation   202
Frank C. Castellana,

Approaches to Modeling Radiocardiographic Data: Comments on F. Castellana's Modeling of the Central Circulation   226
James B. Bassingthwaighte

8 Instrumentation   231
Robert H. Jones, with Leon Lidofsky, Thomas Budinger, and Paul Weber

9 Radiopharmaceuticals in Radiocardiography　254
   Rao Chervu

10 Imaging of Acute Myocardial Infarction　278
   Robert W. Parkey and Frederick J. Bonte

Index　283

Quantitative Nuclear Cardiography

PRINCIPAL EDITOR: RICHARD N. PIERSON JR.

1. Introduction

"In Dreams Begin Responsibilities." quoted by Yeats

Application of radionuclides to the study of the heart has grown with dramatic pace during the past decade. This growth has occurred primarily because the methods were noninvasive and could provide serial measurements over hours, days, or weeks without catheterization. Measurements have been offered well within the central core of the cardiac physiologist's concern: cardiac output, pulmonary blood volume, coronary flow, ejection fraction, end diastolic ventricular volume, intracardiac shunt, and regional ischemia. Some of these measurements have gained full acceptance. Others, perhaps offered prematurely or prior to very basic and far reaching improvements in instrumentation, radiochemistry, and radiopharmacy, require validation for an ultimate appreciation of their relative roles in an arena of competing technologies. The complexities of nuclear medical instruments (particularly during the rapid growth phase when electronic strategies quickly become obsolete) deny most working cardiologists direct and detailed understanding of the inner workings and capacities of these instruments. Many of the components become "black boxes," which for acceptance inherently require an act of faith by the user. Nuclear medicine clinicians have learned to accept their benefits and to absorb their occasional failures, measured in down-time, vague specifications, and qualitative results. The novelty, promise, and magical nature of their techniques have brought this tolerance, as something akin to a mortgage on ultimate acceptance by the clinical community. Cardiologists, unlike many nuclear medicine specialists, accept responsibility for the care of patients who are often critically ill and require frequent and prompt therapeutic decisions. Most cardiologists are disciplined to require quantitative data, with its inherent promise of statistical limits and of representing a continuous function when correlated with a disease state. Thus the normal/abnormal decision is less often used: a left atrial pressure varies from 10 to 20 to 30 to 40 as a *continuous* function of ventricular failure—the higher the pressure the more severe the failure. Arterial pO_2 is not normal or abnormal, but is expressed as 90 or 70 or 50 mm Hg, numerical distinctions which correlate with the disease state. Most of the parameters of cardiology are quantitative, and they are continuous. The additional fact that they are multivariate demands further sophistication in judging numbers. Cardiologists have come to expect and require precise, calibrated, error-limited measurements. Perhaps the most exciting development in nuclear medicine in the 1970s (the authors admit to some prejudice in the matter) is the achievement of a quantitative capability in the recording and interpretation of nuclear medical studies. While the achievement has relevance in many of the specialties served by Nuclear Medicine, in no other field is this achievement more hard won, more demanding of instruments, and potentially more rewarding, than in cardiology.

A great impetus in nuclear cardiography has been the artful mapping of the heart and great vessels by scintiphotography, expanded by timed image montage to outline selected chambers free of the superimpositions caused by the convoluted anatomy of the heart. These studies, done with relatively simple and inexpensive equipment added to the standard gamma-camera, have served notice of the accurate imaging potential of high-dose radionuclide cardiac studies; the subsequent drive has been to provide the quantitation inherently available with radionuclide methods.

> "A Little Knowledge is a Dangerous Thing.
> Drink deep, or taste not the Pierian Spring."   Pope

A strong interaction exists between the adjectives "quantitative" and "digital." One of the important developments of instrumentation which enables Nuclear Medicine to become quantitative is the development of the digital computer to the stage where it is fast enough, small enough, cheap enough, and responsive enough for ordinary mortals to deal with on a friendly basis. Nuclear medicine is inherently bogged down with a vast number of mathematical computations, relating to decay, absorption, nonuniformity, exponential extrapolation, curve fitting, and so on. Properly and conscientiously programmed, the computer can remove the time delays and the tedium from these basic needs, and permit the translation of a complex mass of raw data to a quantitative interpretation within the time frame of clinical decision making at the bedside. For this to occur, the computer, a black box regarded with deep suspicion by conscientious persons in and out of medicine, must be programmed with great care and wisdom, and it must be capable of saying "data inadequate," or "result is $\pm 50\%$," or "my disc drive, and not your patient, has an arrhythmia." Those in the field of computer science have little doubt of the capacity of the computer to revolutionize many fields, including nuclear cardiology. Until the ultimate instrument–computer system is built, a cautious and sceptical attitude is required on the part of the physician wishing to partake of its benefits. No other member of the engineering, technical, or health delivery team can absolve the physician of the responsibility for knowing what went into the logic of the programs which are manipulating data and steering decisions.

It is the goal of this volume to present to the cardiologist, the nuclear physician, the bioengineer, the computer scientist, and the health care systems analyst the details necessary to do quantitative nuclear cardiology in sufficient depth to meet each on his own ground. It is our intention to create scepticism in some, and to decrease it in others, but for all, to define the grounds for doubt and further effort and to indicate, to the extent that solutions are now apparent, the paths which these solutions may take. The Editors envisage a day, and not a distant day, when the cardiologist may routinely, and with no Ph.D. or "computernik" on his staff, say that his decision to give digitalis, diuretic, or pressor drug or refer for catheterization or operation was made on the basis of information deriving from radiocardiography studies. The seventies should be the decade of calibration and validation of these techniques, to the satisfaction of the critical scientist.

This volume is intended as a professional text rather than a compilation of papers presenting the state of the art. Organization of the material has evolved greatly with study of the components. The final format is addressed largely to method and concept, recognizing that instrumentation, technique, and concept are deeply intertwined. Most information is available on instruments and radiopharmaceuticals, fields in a rapid state of beneficial flux, in that the new surpass the old. Even in the less explored areas of the disciplines, some principles are firm, and they can be delineated. Less is firmly known about the ability of these methods to add to or replace traditional angiography and catheterization techniques: too few patient studies have been reported and cross-compared. The clinical intersection of nuclear medicine and cardiology will vary greatly from one institution to another, depending on the available strengths of the measurement disciplines, the patient material, and to some extent, on "who got there first." It seems valuable to the editors to reduce to writing even at this early time, when chaos still abounds, the principles which have emerged. We hope that this has been done in such a way that future toilers in the vineyard may avoid some of the errors of ignorance that have waylaid us, in the search for better methods. The complexity of the multidiscipline, as well as rapid growth and change, have been responsible for most of these errors. More effective collaboration at discipline interfaces is now occurring at many centers. Most of the solutions envisaged here, for example application of the mathematical model of the circulation to patient studies, will require further extensive collaboration during the research and development phase, although the goal of such efforts remains a developed system.

Our text treats nuclear cardiology as a series of disease states for which radionuclide diagnostic strategies exist. The methods are subordinated to the clinical purposes for which they are used. This organization submerges the identities of individual authors and editors in favor of a cohesive disease oriented presentation. This choice of format is directed to the cardiologists who will ultimately define the reality of nuclear cardiology as a contributing element in patient care. We believe that the other partners in the search, the practitioner of nuclear medicine and the radiopharmacist, engineer, and computer specialist can find their way in this structure, and that they will understand that the aim of this book and these efforts it to serve the cardiologist and his patients.

PRINCIPAL EDITOR: WILLIAM J. MACINTYRE

CONTRIBUTING EDITOR: JOSEPH P. KRISS

2. The Normal Heart

The recent development of methods using scintillation cameras capable of recording sequential radionuclide images from the precordium has provided a means of visualization of both anatomic and functional characteristics of the heart and great vessels. This has been accomplished to bypass the hazards of cardiac catheterization, or rapid administration of large volumes of radiopaque dye into the circulation under high pressure, as is necessitated by radiographic methods. By use of this methodology, scintigraphic images produced by gamma-emitting radionuclides are recorded in rapid sequence as the isotope flows through the circulatory system. These images can then be analyzed in two ways. First, the time sequence of chamber filling and the position and size of the chambers and vessels can be demonstrated in a manner similar to contrast angiography, a qualitative technique which permits the interpretation of the sequential dynamic images as representations of pathophysiologic phenomena. Success in visualizing and identifying specific cardiac chambers and great vessels, and recognition of the quantitative nature of radionuclide studies, have greatly stimulated attempts to measure various parameters of heart function (1–6).

Second, since isolated anatomic regions of the central circulatory system can be identified by external imaging, relatively pure dilution curves can be obtained from those selected regions. Thus, the problems of measuring transit time between selected vessels or chambers have been considerably simplified as compared with the original single-probe measurements in which the dilution curves obtained were spatially integrated over the entire heart.

Previous methods to obtain these measurements with similar accuracy required cardiac catheterization. Full quantitative analysis of the available data in a radionuclide angiocardiogram now permits the measurement of left and right ventricular ejection fraction, stroke volume, end systolic and end diastolic ventricular volumes, pulmonary blood volume, cardiac output, and the magnitude of shunts at identifiable anatomic levels. Proper utilization of this technic requires a complete definition of the ranges of normal.

Positioning Technique

Adults may be studied in the supine or sitting positions. The anterior position is often routinely used. For this examination the detecting head is positioned close over the precordium, centered 1 to 2 in. left of midline. The heart should be centered in the field of view, with some lung field visible. This alignment may be achieved with precision by use of a transmission source placed behind the patient. However, the anterior view is suboptimal for visualization of the region of the left atrium because of overlying heart structures. Better delineation of the left atrium and the left ventricle can be accomplished by centering the detector head over the left anterior axillary line in the 30° left anterior oblique (LAO) position with the face of the detector directed caudad at an angle of about 15°. This "modified LAO" view permits visualization of the left atrium relatively free of overlying cardiac structures. In the absence of malrotation of the heart or marked right heart enlargement, the interventricular septum is orthogonal to the plane of the camera head, and is commonly identified in this view, and the left and right ventricular chambers are thus separated. This view is of value for assessment of chamber size and myocardial wall thickness.

In measuring left ventricular volume or left ventricular ejection fraction by methods depending on planimetry of area, as in cineangiography, the conventional right anterior oblique (RAO) or left lateral positions are preferred by some investigators, based on the assumption that the contents of the left ventricle in diastole form the outer borders of the cardiac image in these views.

Infants and very young children should be studied supine, as special problems exist. Crying is inevitable in the unsedated child, and since unpredictable movement may invalidate the study, sedation is recommended. The circulation times tend to be be very rapid and the size of the heart and the resulting scintiphoto images are very small. A converging collimator will be helpful in enlarging the image two- to threefold. Since the resulting enlargement increases with increasing distance from the center of the collimator, the resulting image cannot be analyzed by area–volume methods without complex corrections, but these measurements are not usually desired in pediatric studies, and the use of the converging collimator offers great merit. In addition, the pinhole collimator has also been used for enlargement (7).

Injection Technique

It is highly desirable that the radionuclide enter the heart as a compact bolus. While satisfactory studies may be obtained by intravenous injections using special syringes, flushing, hand pumping, tourniquet removal (8), or arm elevation, it is preferable when possible to assure right atrial input of nondispersed tracer by injecting via a central venous catheter with its end lying in the superior vena cava. If detailed analysis of chamber volumes by mathematical analysis of dilution curves is intended, as outlined in Chapter 7, an EKG-triggered, powered injection into the right atrium during systole may be utilized. Failure to deliver a bolus rapidly may lead to poor delineation of anatomic structures and erroneous interpretation regarding valvular disease, shunts, or left ventricular performance; the errors are increasingly severe in each downstream chamber.

Many different radiopharmaceuticals could be used, depending on the imaging equipment available, the need to repeat studies serially, and available radiopharmaceutical skills. Most investigators have used ^{99m}Tc as pertechnetate or tagged to human serum albumin (TcHSA), because of the high count rate available and the ease of collimation of the 140-keV gamma ray. If cardiac output and other quantitative methods are to be applied, a tracer which remains within the blood volume must be used. In Chapter 9 tagging methodology and other techniques are discussed, including serial stepup in gamma energy so that repeat studies can be done over a short time period. With ^{99m}Tc, the usual adult dose is 10–20 mCi; in children 0.1 mCi/kg has been suggested; Table 1 indicates appropriate dosimetric considerations. However, it must be recognized that these studies are rarely carried out in normal subjects and that the relative cost in radiation absorbed must be compared with traditional angiographic methods rather than the absolute magnitude of the radiation to the patient in determining the radionuclide dose; cardiac catheterization rarely involves less than 5 rads, and 25 rads is often delivered in seeking comparable diagnostic detail.

Table 1[a]. Absorbed Radiation from ^{99m}Tc-Albumin

Age	Whole Body Weight (g)	Whole Body Dose from ^{99m}Tc-Alb (rads/mCi)	Dose to Critical Organ (rads/mCi)	
			From TcO$_4^-$ (gut)	From Tc-Alb (blood)
New born	3,540	0.2	1.6	1.0
1 year	12,100	0.06	0.4	0.3
5 years	20,300	0.04	0.3	0.2
10 years	33,500	0.03	0.2	0.1
15 years	55,000	0.02	0.1	0.06
Standard man	70,000	0.02	0.1	0.06

[a] Adapted from Cloutier and Watson.

DESCRIPTION OF ANATOMICAL FEATURES

Figure 1 demonstrates the angiographic findings typical of the normal subject studied in the anterior position (9). In this and subsequent figures, the time interval in seconds after injection is shown for each scintiphoto frame. The following abbreviations are used in this and subsequent figures: SVC, superior vena cava; RA, right atrium; RV, right ventricle; PA, pulmonary artery; RL, right lung; LL, left lung; LA, left atrium; LV, left ventricle; A, aorta.

The phase of right-heart filling is seen in the first frame. In the second frame the right ventricle, pulmonary artery, and initial lung activity are noted. Following intravenous injection, the flow of radioactivity from the superior vena cava through

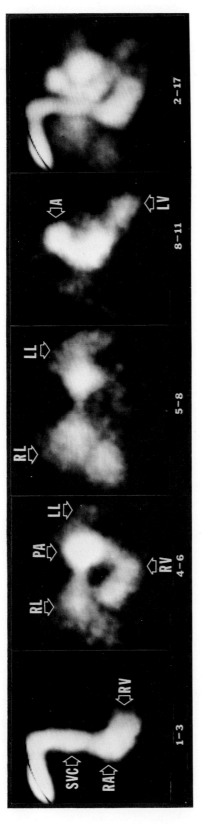

Fig. 1. Radionuclide angiocardiogram of normal subject, anterior view. Time interval in seconds after injection is shown for each scintiphoto frame. The phase of right-heart filling is seen in the first frame (SVC, superior vena cava; RA, right atrium; and RV, right ventricle). In the second frame the right ventricle, pulmonary artery (PA), and initial lung activity (RL, LL) are noted. Frame 3 shows activity primarily in lung. Frame 4 shows left ventricle (LV) and ascending aorta (A). Frame 5 shows the summed images from 2–17 seconds.

the right atrium and right ventricle to the pulmonary artery usually lasts less than 6 seconds and describes a typical U-shaped pattern as seen in these first two frames.

Lung activity is seen most clearly in frame 3. As shown in this frame, during the time of pulmonary transit of radioactivity there is little intracardiac tracer for a few seconds if there has been satisfactory bolus injection. Left atrial and left ventricular filling are commonly noted by 8 to 15 seconds and in this study the levo phase, delineating the left ventricle and the ascending aorta, is noted between 8 and 11 seconds in frame 4.

The left atrium is not normally well visualized in the anterior view because of superimposed or adjacent activity in the left ventricular outflow tract and ascending aorta. The modified LAO (30° oblique, 15° caudad tilt) view is preferred for optimal left atrial visualization. The arching course of the left pulmonary artery is useful as a landmark, for the left atrium subsequently fills just beneath it.

The normal left ventricular cavity appears as an elliptical shape, sometimes tapering at its distal extent, and has a width usually not exceeding that of the normal aortic root (frame 4). The region of the aortic valve may or may not be clearly seen in the anterior view and the descending aorta is not as well separated from other chambers in this position as in the LAO view.

A composite play-through of the complete cardiopulmonary circulation phase from 2 through 17 seconds is pictured in Figure 2. Also shown in this figure are dilution curves recorded from the regions of right ventricle and left ventricle as designated by the cursors shown on the composite photograph. Note the contribution of the right ventricle to the dilution curve recorded from the left ventricle site and vice versa. This contribution has been termed "cross-talk" or "contaminating counts"; it is dealt with explicitly in Chapters 5 and 6.

Greater separation of the two sides of the heart can be accomplished by the LAO view previously described in which the camera face is placed perpendicular to the interventricular septum. A comparison of the anterior view with the LAO view is shown in a second normal subject in Figure 3.

In this figure it is noted that on frame 3 the pulmonary venous filling overlaps the region of the left atrium, designated as X, in both views. The oblique position is usually preferred to visualize the left atrium (frame 4), and the interventricular septum shown in the lower composite on frame 5.

Verification of the successful separation of the right and left ventricles is demonstrated in Figure 4 in which no contribution from the other ventricle is seen on either dilution curve recorded from the regions between the cursors.

QUANTITATIVE OR DIGITAL RECORDING

A different type of information can be gained from the same data if they are recorded with appropriate care in such a way that count rate is related to tracer concentration. For this purpose, it is essential to select carefully the sites or "regions of interest" from which recording is carried out. For this need, a method of replaying the original data is required. Various techniques are covered in Chapter 8. As shown by Figures 2 and 4 this objective has been accomplished by visually recognizing the various structures as previously described, designating certain regions on

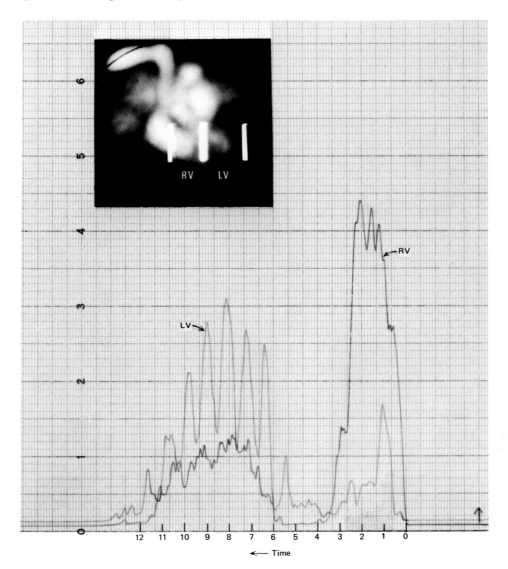

Fig. 2. Upper left image demonstrates a composite play through of the Cardiopulmonary circulation phase frame from 2 through 17 seconds. Solid white lines represent cursors designating regions (right ventricle and left ventricle) from which the dilution curves were drawn. Note contribution of right ventricle to left-ventricle curve and vice versa. Left-ventricle curves are shown to resolve individual contractions of the heart.

the camera field of view as characteristic of an individual chamber (flagging) and plotting the relative count rates in the flagged area at short time intervals (frames) as a function of time.

Although originally proposed as an analog system (10), the need for such refinements as dead time adjustment, nonuniformity correction, background subtraction, and data manipulation has indicated the advantages of digital reduction of data, and therefore computer processing.

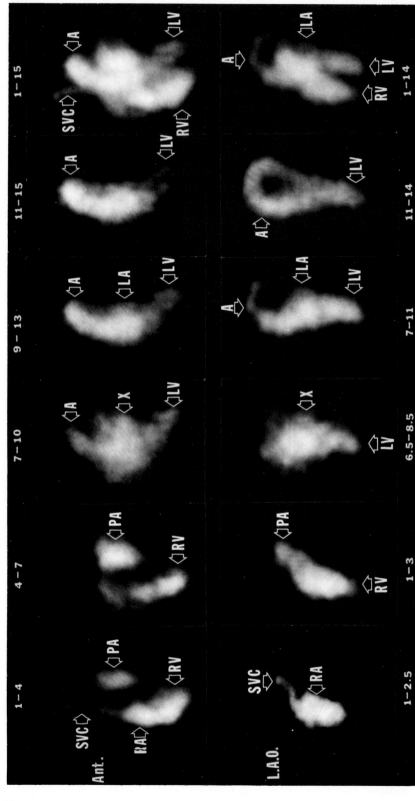

Fig. 3. Comparison of normal radionuclide angiocardiogram recorded in both the anterior (upper figures) and the left anterior oblique positions, LAO (lower figures). Note resolution of the intraventricular septum in LAO frame 5. Nomenclature similar to that in Figure 1 except that X in frame 3 designates the overlap of pulmonary venous filling with the left atrium. The oblique position is usually preferred to visualize the left atrium (frame 4).

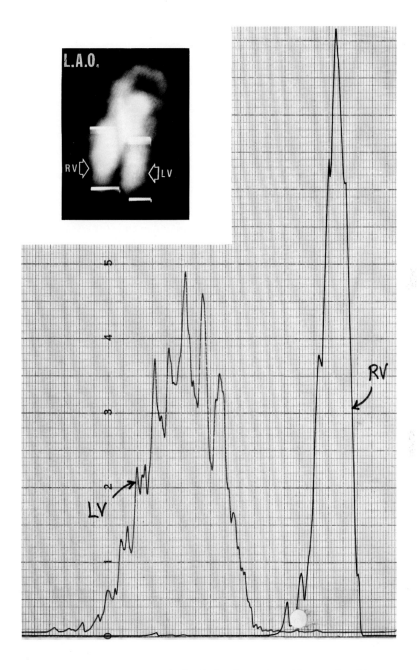

Fig. 4. Composite of central circulation from 1 to 14 seconds with cursors designating areas of left and right ventricles from which the dilution curves were drawn. Note the freedom from cross talk from the separate chambers.

Such studies have been performed some time ago with both the multicrystal autofluoroscope (11) and the Anger scintillation camera (12). In general, the technique involves conversion of the analog positioning signals or individual crystal locations to a digital location in a memory core, and transfer of each successive frame to a digital disc or tape recorder for recall and computer processing. Achieving statistical data density and handling the data stream sufficiently rapidly represent the instrumental limits imposed on this approach. Chapter 8 provides a detailed review of available methods. Linearity of instrument response is a basic requirement for quantitative analysis over 12,000 to 15,000 cps (counts per second); most gamma-cameras lose this linearity because of long dead-times. This problem is described in detail in Chapter 8. The remaining material in this chapter has been recorded with instruments corrected for linearity.

In the following examples data are accumulated on a 1600-channel analyzer as a digitized 40×40 matrix in a pattern corresponding to the isotope location within the body. Since the diameter of the collimator is 10.5 in., each element of the 40×40 matrix can be represented spatially as a square with sides approximately one-quarter inch in length, which is about the optimal resolution of the camera system with collimator (Chapter 8).

As the isotope bolus flows through the heart and great vessels, the rapidly changing sequence of images is recorded at 0.6-second intervals and transferred to a digital tape recorder. Data are then printed out for digital computer processing in the form of 40×40 elements for each frame sequentially. Regions of interest are identified on these printouts by spatial and temporal analysis. The essential elements of the data acquisition and retrieval system described are shown diagrammatically in Figure 5 (13). Recording of frames at 0.9-second intervals, with two thirds of available data collected, is appropriate to "medium frequency" studies, useful for transit time analysis. Newer and faster systems capable of forming frames as rapidly as 30, 50, or 100 frames/second have been developed for high-frequency data analysis suitable for study of ventricular contraction and relaxation. These studies invoke new sets of problems and limitations and they are considered separately under Ventricular Function in Chapter 5.

A sequence of printouts of individual matrices during the rapid passage of the $^{99m}TcO_4$ through the central circulation is shown in Figure 6. (14) Various symbols are printed out to represent the contours at cut-off levels of 25, 40, 60, and 80%. In frame A at the top left the ^{99m}Tc-pertechnetate has just arrived at the superior vena cava. In frame B the injected material has progressed into the chambers of the right heart and by frame C some activity at the upper right shows a progression of the radioactivity from the right ventricle out to the pulmonary artery. Each frame represents a collection time of 0.6 seconds plus a transfer time of 0.3 seconds.

By frame D (7.2 seconds after injection) the pertechnetate is fairly well distributed in the lung circulation and frames E and F (14.5 seconds) show the return of the radioactive material from the pulmonary circulation back to the chambers of the left heart. In frame E there is some residual radioactivity still clearing from the lung circulation, but these counts are largely under the 25% contour level.

Frames A, B, and C can be lumped together as shown in Figure 7 A so that the composite represents the right side of the heart. The actual counts accumulated during these three frames are shown by the matrix on the upper left and the symbols depicting the various level contours are shown on the lower left.

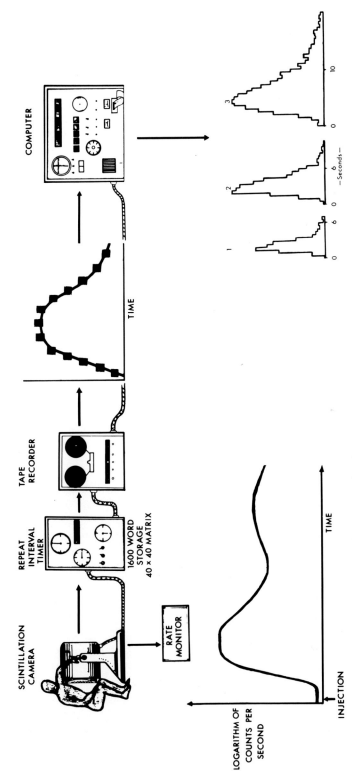

Fig. 5. Schematic diagram of data acquisition and retrieval system.

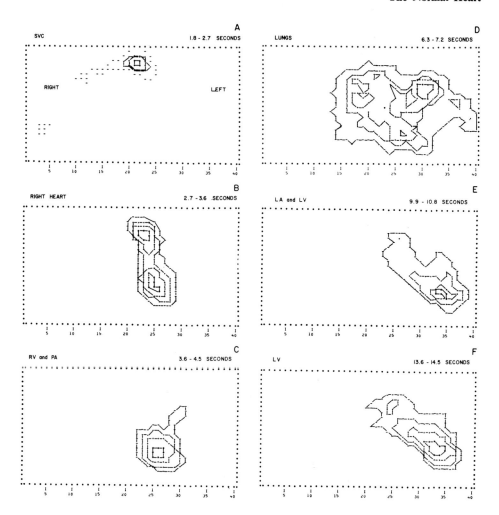

Fig. 6. Composite of six contour plots reflecting the passage of $^{99m}TcO_4^-$ through the central circulation. Upper left is frame 3 (A) (2.7 seconds) shows appearance of material in the superior vena cava. Middle and bottom frames on the left were recorded at 0.9-second intervals following the initial appearance. Top right is frame 8 (D) (7.2 seconds) shows distribution in lung circulation. The following two frames are 12 and 14 and show return to left chambers of the heart in the 10.8- to 12.6-second range.

The left side of the heart can be similarly delineated by lumping frames from 9.9 to 13.5 seconds (Fig. 7 B). Inclusion of frames after this point would reflect radioactivity in the ascending and descending aorta, while distributions somewhat earlier than frame E would include residual radioactivity in the lungs.

Finally, a composite of both sides of the heart is illustrated in Figure 8. Again the actual number of counts collected in each element is shown on the top and the contour symbols on the bottom. Note that the symbols denoting the 25 and 40% contour levels are close together, usually no more than one element apart, demonstrating that the fall off in count rate at the border of the heart cavity is quite steep, giving confidence of a digital nature to the localization in two dimensions of the projection of the heart on the camera crystal. It is important to observe that 400–600 counts/cm^2 define this area, permitting reasonable statistical resolution.

The three dimensional views of the matrices in Figure 7 are shown in Figure 9. The upper figures represent the right heart with lumped frames 3–5, the middle view shows the lumped frames 11–15, and the lower view shows the summation of the two. The two views illustrated are taken from the heart with left and right projections. Two similar projections from the apex of the heart looking towards the base are also recorded but are not shown here. These representations thus provide the location of the various chambers in the matrix field so that dilution curves characteristic of specific chambers or vessels can be obtained.

In order that dilution curves that are derived from one chamber alone and, thus, represent a single concentration* curve may be obtained from the sequential printouts, it is desirable to select the region of interest from that part of the compartment that does not superimpose on any other chamber. If this is not possible, it is desirable then to select a region where the superimposition of counting rate is most widely separated in time so that cross-talk contamination between compartments can be more easily identified and separated. For utilization of the camera to study a dynamic process, both spatial and temporal superimposition must be considered, and identification of specific compartments must be attempted by obtaining optimal resolution in both domains. An example of the selection of regions of interest from the matrix printouts is shown in Figure 10 (15). The illustrated regions are rectangular (although they may be selected in any shape) and are chosen to exclude ambiguous areas which may overlap chambers.

After the identification of each region of interest, time–activity curves from sequential matrices are read out from areas representing the superior vena cava, right atrium, right ventricle, a lung field, left atrium, left ventricle, and aorta. These curves can be read out in the form of an incremental histogram for a representative number of regions as is illustrated in Figure 11.

ANALYSIS OF DILUTION CURVES

The use of camera techniques and the identification of specific regions have given a mechanism for multirecording of specific time–activity relationships that should allow both transit and volume parameters to be derived for all the compartments.

Calculation of Mean Transit Time

In a vascular section in which there are no stagnant pools and in which complete mixing occurs, the mean circulation time may be expressed as the volume of blood between the injection and sampling sites divided by the blood flow rate. The uncertainty of the measurement is dependent on the length and complexity of the vascular section and the degree of spread in the distribution of transit times, as well as exact identification of the injection and sampling sites.

---

* See Chapter 6, p. 223, for an explicit discussion of the relative contributions of tracer mass, tracer concentration, and cross-talk to this concentration curve.

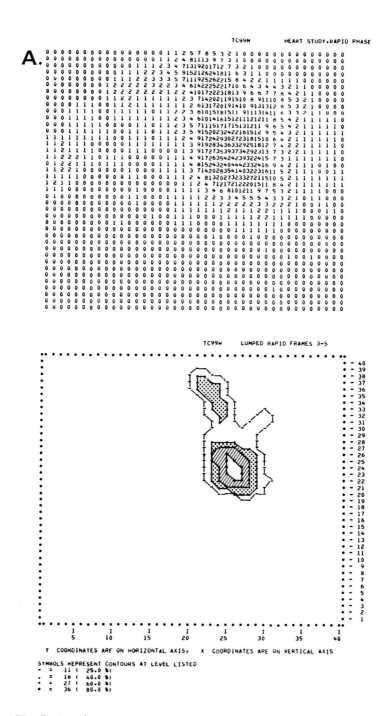

Fig. 7. Matrices and contour symbols obtained from lumping frames 3, 4, and 5 to show right heart areas (*A*) and frames 11 through 15 to show left heart areas (*B*).

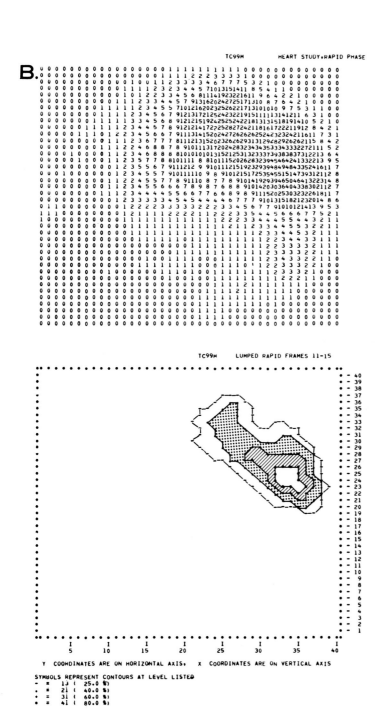

Fig. 7.—Continued

TC99M   HEART STUDY, RAPID PHASE

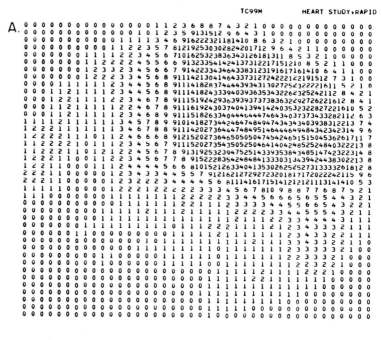

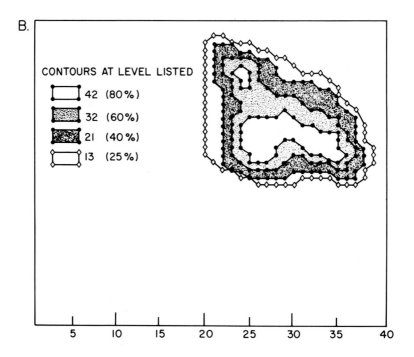

Fig. 8. Matrix (*A*) and contours (*B*) of heart outline obtained by combining matrices of Figure 6*A*, omitting frames when most of the activity is in the lung field.

Analysis of Dilution Curves

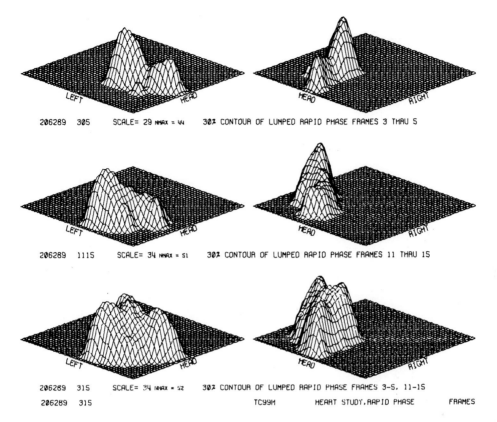

Fig. 9. Three-dimensional representation of the matrices in previous figures. Top figures show relative deposition within the lowest contour border of the right heart of Figure 6a. Middle figures show the similar representation of the left heart in Figure 6b. The bottom figures show the combined heart outline similar to Figure 7.

Provided the injection closely approximates an instantaneous delivery, the mean transit time (MTT) may be expressed as the centroid of the indicator dilution curve through that section.

Thus:

$$\mathrm{MTT} = \frac{\int_0^\infty t\, c(t)\, dt}{\int_0^\infty c(t)\, dt} \tag{1}$$

Although the departure of the injection from an instantaneous delivery will influence the mean transit time of the first vascular section, the mean transit times of the sequential chambers or sections may be determined as the difference in mean transit times from sequential regions.

It is this feature that has made the multirecording of dilution curves from various sites advantageous in the study of intracardiac circulation times. In a study using the digital autofluoroscope described in Chapter 8, radionuclide angiocardiograms were obtained in normal subjects.

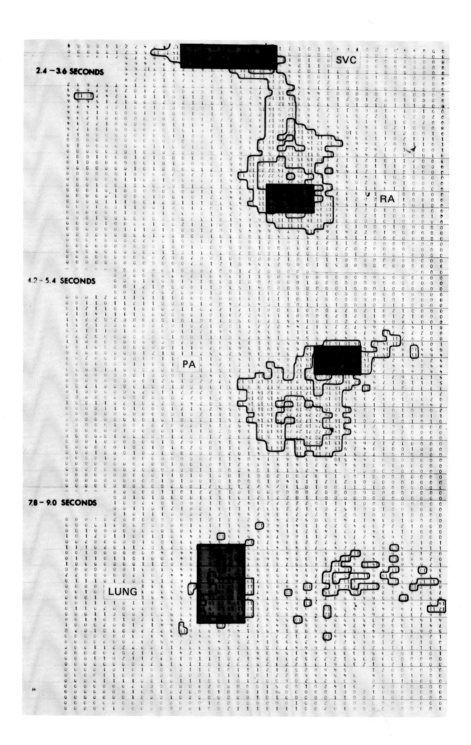

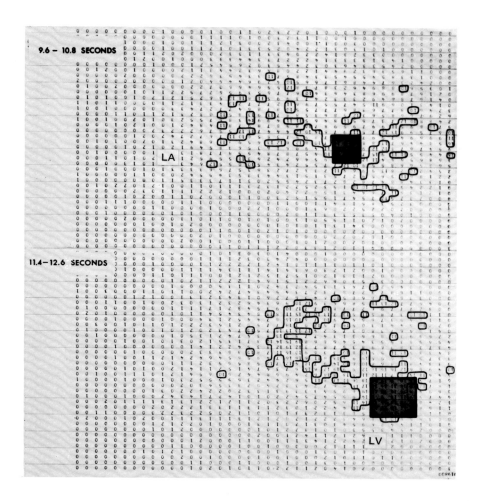

Fig. 10. Computer printout with contour lines showing areas of increased counting rate and shaded areas representing the regions of maximum counting rate which are flagged for derivation of dilution curves for specific compartments. When regions superimpose, as with the right atrium (RA) and right ventricle (RV) in this example, it is desirable to select areas most widely separated in space and time. In this case, selection was made of the region of outflow of the right ventricle to the pulmonary artery for best separation of RV from RA.

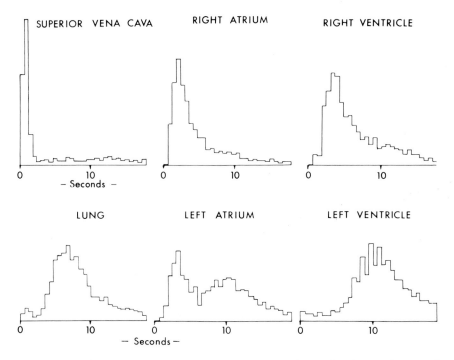

Fig. 11. Dilution curves recorded from the six sites selected in Figure 10. Note the contribution of right heart or pulmonary artery to site selected for left atrium.

Counts from each detector of the 294 crystal matrix were accumulated at 0.2-second intervals during the study, and an additional 0.03-second delay was required to record data onto computer tape. Therefore, each 0.2-second counting interval began 0.23 second after the preceding one. The detector was positioned in contact with the chest directly anterior to the precordium of resting, supine patients. A 10-mCi bolus of ^{99m}Tc-pertechnetate (or 0.3 mCi/kg in children) in a volume less than 2 ml was rapidly injected into a large arm vein through a 19-gauge needle. Patients with an indwelling intravenous catheter at time of study received injection through the catheter with an immediate flush of saline solution (16).

Appearance and Transit Times

Total counts for each accumulation interval of the study were obtained from detector groups corresponding to the right atrium, right ventricle, pulmonary artery, right and left lungs, left atrium, left ventricle, and aortic arch. Grouping of detector units with similar fluctuations of radioactivity permitted recognition of discrete cardiac areas in all patients studied. Semilogarithmic extrapolation of each regional indicator-dilution curve permitted calculation of mean transit times by Equation 1.

The delineation of these areas on the detector matrix is shown in Figure 12. The time of maximum count rate of the primary and secondary component curves determined for each detector unit depicts the flow of radioactivity over the detector field. Timing of the passage of tracer bolus through the central circulation by the primary component permits recognition of these anatomic regions with sufficient definition to identify the right and left heart chambers and to separate the aortic arch and pulmonary outflow tract. The time of peak count rate of the second component curve for each data point further enhances anatomic resolution. Along the cardiac septum, secondary curve components are identified within the primary right-heart region which contains left-heart counts, and a zone of detector units within the left heart chambers demonstrates a secondary component with a time of secondary peak count rate that corresponds to the right-heart peak time. In addition, the aortic arch region contains a secondary component of radioactivity arising from the pulmonary outflow tract, and vice versa.

In Figure 13 characteristic curves from cardiac regions outlined by the time course of radioactivity in a normal subject demonstrate sharp appearance and passage of the radionuclide bolus through the right heart chambers. Superimposition of counts from the pulmonary artery and left atrium are seen to produce biphasic curves from these regions. Radioactivity from the pulmonary outflow tract is also seen included in data from the aortic arch. However, components of these complex curves are usually sufficiently well defined to permit separation by extrapolation for calculations of transit time. Incremental passage of tracer through the heart is particularly apparent in data from the left ventricle where fluctuations of count rate with systole and diastole are clearly seen if the time frames are of sufficiently high frequency (10 to 50 frames/second).

As shown in Figure 14, 10 normal subjects demonstrate similar times of radionuclide transit through the central circulation.

In the upper left the heart model, A, shows the average time of appearance of the tracer in the right atrium, right ventricle, pulmonary artery, lung, left atrium, left ventricle, and aorta. Very little delay in appearance is seen between right atrium and right ventricle or between left atrium and left ventricle. As would be expected, the most noticeable delay occurs between the right ventricle and left atrium because of pulmonary transit of blood. Here it is seen that the tracer first appeared in the left atrium 5.1 seconds after appearance in the right atrium.

Model B shows the time of peak count rate of the curve from each site. In normal subjects, the radionuclide bolus retains a rather constant configuration during passage through the central circulation, and peak times demonstrate a consistent relation to mean transit times. Peak times offer a simple index of blood transit, but are less reliable for actual measurement of blood flow than mean transit times, which are not influenced by alteration of bolus configuration.

The mean transit time for each site is depicted in model C. These values have provided the most reproducible index of blood transit. In these 10 normal subjects, mean transit time from the right atrium to the left ventricle averaged 9.2 ± 1.2 seconds and pulmonary mean transit time averaged 6.6 ± 1.1 seconds. Use of similar injection techniques has produced cardiac chamber mean transit times with standard deviations ranging from 0.5 to 1.6 seconds.

It is important in these determinations to give careful attention to the images of count-rate density to avoid errors arising from inaccurate anatomic assignment of counts. The magnitude of data alteration which may result from minor errors in

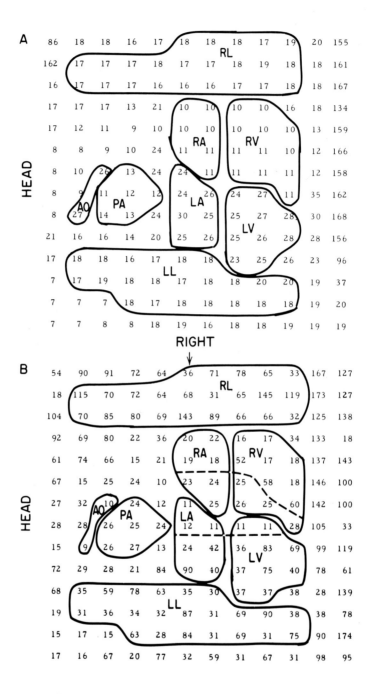

Fig. 12. Anatomic resolution obtained by computer identification of the time response of each detector. Each time unit represents 0.23 second. (A) The time of maximum of the primary component curve delineates cardiac regions. Spatial resolution achieved is well illustrated by separation of the right ventricle (RV) from the right lung (RL), left ventricle (LV), and liver, and by separation of the pul-

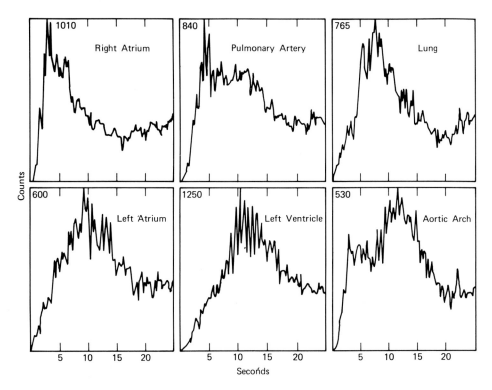

Fig. 13. Radionuclide angiocardiogram demonstrates curves typical for normal subjects. Biphasic curves are recorded from regions superimposed anatomically, such as the pulmonary artery and left atrium, but these curves retain sufficient definition to permit separation into two components. Time resolution is sufficiently discrete to define cardiac contractions, particularly in the left ventricle.

assignment of areas can prove quite great. Comparison of radionuclide angiocardiogram data from correctly assigned cardiac areas with data obtained by moving the same areas only 1 cm to the left demonstrates severe curve distortion (Fig. 15). All dynamic radionuclide determinations of regional organ function which cannot be oriented to anatomic structure by a consistent, reproducible method must be considered potentially highly inaccurate. It is important to realize that nothing in the shape of the curve recorded indicates that an error in flagging the region of interest as left ventricle, pulmonary artery, and so on has been made. Thus meticulous care must be taken to identify the projections of anatomic landmarks on the imaging screen prior to identifying regions. For purposes of mean-transit-time analysis, smaller areas well within the chamber boundaries are selected for mini-

---

monary artery (PA) and aortic arch (AO). (*B*) The time of maximum of secondary data contributed to each detector from adjacent anatomic areas further defines cardiac chambers. Detectors along the cardiac septum (dotted lines) record primary data from the right or left side of the heart and secondary data from the other.

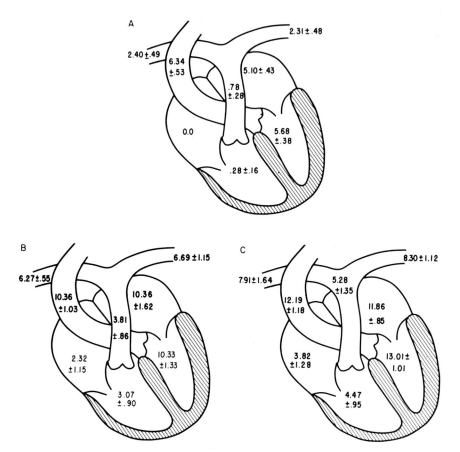

Fig. 14. Blood transit through the central circulation is summarized for 10 normal subjects. (*A*) Appearance times maximum. Delay in appearance occurs between the right ventricle and left atrium because of pulmonary transit of blood. (*B*) Peak times. (*C*) Mean transit times which provide the most reproducible index of blood transit.

mizing risk of contaminating count contributions from adjacent chambers. In fact the region of interest should be as small as possible, the lower limit being set by a need for statistically acceptable count rates. Areas comprising one quarter to one half of the left ventricle provide count rates in the range of 500 to 1000 per 0.1 second at peak, and 50–100 after washout, when doses in the range of 5–10 mCi of ^{99m}Tc-albumin are used. The assumption of complete mixing of the injected tracer in the chamber where measurements are made is central to this technique. The assumption is widely, but not universally, accepted.

Application to Chamber Volume Measurements

The mean transit time of a vascular section is dependent on the volume between the injection site, and the flow through the system. Since it is possible to derive the

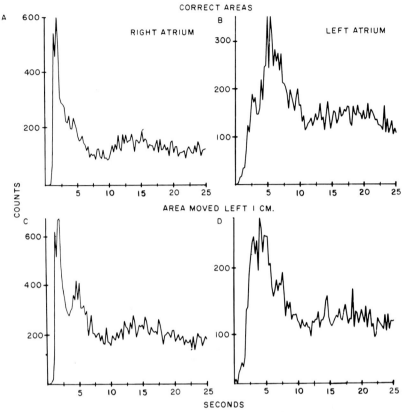

Fig. 15. The influence of anatomic region selection on the accuracy of dynamic studies is illustrated by this radionuclide angiocardiogram data. Only a 1-cm movement of the properly selected atrial regions greatly alters the curves produced.

mean transit time from analysis of the indicator dilution curve, it should also be possible to obtain the volume of the vascular segment provided the flow through the system is known.

This has been accomplished by utilizing the sequential dilution curves of Figure 11 to be analyzed by the analog computer fitting analysis shown in Figure 16. Each compartmental unit may be represented as an input–output relationship and may thus be used to construct an analog computer model expressed simply in terms of the Laplace transform and transfer function. By searching iteratively for the closest fit of the model curve to the data curve, the property of each compartment can be determined in terms of flow/volume (T_i) or transport delay (τ_i) on the computer controls. This fitting procedure is performed step by step in the order of the course of the tracer by cascading the determined output onto the next compartment as an input.

When adjusted for best fit, the computer parameters yield the time constants, T_{SVC} for superior vena cava, T_{RA} for right atrium, T_{RV} for right ventricle, T_L for

ANALOG SIMULATION METHOD

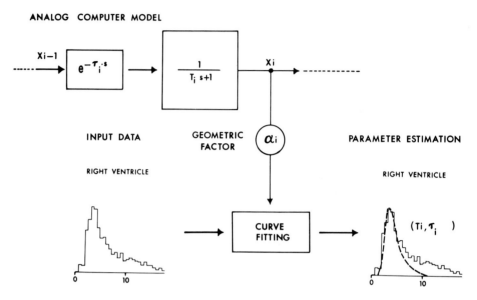

Fig. 16. Block diagram representing one compartmental unit which reconstructs a simulated dilution curve of the compartment. By fitting the simulation curve on the right to the input data on the left, parameter estimation of the compartment can be accomplished. The block consists of a first-order delay system and simple time delay.

the mixing part of lung, T_{LA} for left atrium, and T_{LV} for left ventricle, and τ_p for the time delay part of the lung. If the cardiac output or flow rate, F, is determined, the distribution volume, V_i, defined for each mixing chamber can be expressed as:

$$V_i = T_i \times F \qquad (2)$$

Since the lung has a delay component, τ_p, the pulmonary blood volume, PBV, can be expressed as:

$$\text{PBV} = V_p + F \times \tau_p \qquad (3)$$

If a dilution curve is contaminated by a dilution curve from another compartment (as is usually the case for the curve for the left atrium, which is frequently obscured by activity from the outflow tract of the right ventricle), the analog fitting analysis requires that a single compartment curve be derived by subtraction of the undesired component from the composite curve.

An illustrated example of curves derived from the regions of interest selected from the sequential printout of each frame is shown in Figure 17. The simulated curve is the dotted line superimposed over the recorded histogram. As a first step, the exponentially extrapolated area under each dilution curve was measured by a planimeter. This area ratio was set on the computer control as the geometric factor, α_i, before the fitting procedure was started. The output from the computer for each dilution process was fit to the recorded dilution curve by iterative adjustment of

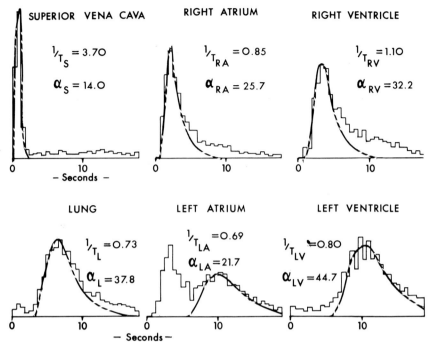

Fig. 17. Results of analog simulation of curves derived from regions designated in Figure 12. Broken lines represent the simulated analog curves. Sequential procedure of fitting yields potentiometer reading from the computer as $1/T$ and τ values.

the T_i values on the computer control in the order of the course of tracer bolus through the circulatory system. In the illustrated example, the injection input was approximated by a rectangular wave input with a time duration of 0.7 second. Following this input, the time constant of each chamber was selected sequentially in this order. There is a time delay between right ventricle and lung, and lung and letf atrium, which in this example amounts to 4.0 seconds. For each fitting step for selection of T_i values, two or more iterations were required. On the average, all these procedures required about 15 minutes for completion.

The volume of each heart chamber of a series of vascular segments in the central circulatory system was calculated according to Equation 2, and pulmonary blood volume was calculated according to Equation 3, after determination of the chamber parameter by analog computer fitting analysis. Table 2 shows the values for the volumes of each compartment for 9 patients without hemodynamic abnormalities.

Current Status of These Methods

Clinical application of cardiac chamber scintiphotography has been pursued on a large scale by Kriss at Stanford, and elsewhere. The patterns described here are obtainable with a standard gamma camera and $^{99m}TcO_4$ pertechnetate studies, requiring only a means of rapid sequence imaging or of storing data for replay, or both. This equipment is widely available. Interpreting the patterns requires sub-

Table 2.

Patients	Diagnosis	Cl (liters/ minute/ m²)	SVC (ml/m²)	RQ (ml/m²)	RV (ml/m²)	PBV (ml/m²)	LA (ml/m²)	LV (ml/m²)
1	Dermatitis	4.13	—	139	125	270	112	71
2	Dermatitis	5.55	38	83	75	383	68	57
3	Hypertension	3.90	18	78	59	345	96	83
4	Hypertension	6.00	13	71	108	425	99	89
5	IHD	2.58	17	72	103	265	70	85
6	Pyelitis	3.80	18	85	59	290	132	
7	Pulmonary fibrosis	3.18	21	101	77	285	83	81
8	Pulmonary embolism	2.76	42	94	98	230	64	105
9	Hypertension	3.94	62	130	81	243	112	78
	Mean		28.6	94.8	88.3	304.0	88.0	81.1

stantial experience; the close cooperation of a cardiologist–angiographer, both as a source of referral of case material and as a guide to the anatomy of the heart, will speed the day when the ultimate utility of this method can be evaluated. The low patient radiation dose and noninvasive character of the study should provide a large diagnostic arena in outpatient as well as inpatient study if the specificity and sensitivity of the technique can be shown in other laboratories to be as accurate as here indicated.

Acceptance by cardiologists of the transit time/volume data described remains to be assessed. A generation of radiocardiographic work by Donato and his successors (see Chapter 5) using single probes suffered from severe limitations in spatial resolution, low count rates, and incomplete mathematical analysis of the circulatory system. The tools are currently at hand for a new study of the range and efficiency of quantitative flow/volume measurements in a clinical arena. Such studies have not yet appeared.

REFERENCES

1. D. T. Mason, W. L. Ashburn, J. C. Harbert, L. S. Cohen, and E. Braunwald: Rapid Sequential Visualization of the Heart and Great Vessels in Man Using the Wide-Field Anger Scintillation Camera, *Circulation* **39:** 19–28, 1969.
1. G. Burke, A. Halko, D. Goldberg: Dynamic Clinical Studies with Radioisotopes and the Scintillation Camera: IV. ^{99m}Tc-Sodium Pertechnetate Cardiac Blood-Flow Studies, *J. Nucl. Med.* **10:** 270–280, 1969.
3. J. P. Kriss, and P. Matin: Diagnosis of Congenital and Acquired Cardiovascular Diseases by Radioisotopic Angiocardiography, *Trans. Assoc. Am. Physicians,* **82:** 109, 1970.
4. P. Matin, and J. P. Kriss: Radioisotopic Angiocardiography: Findings in Mitral Stenosis and Mitral Insufficiency, *J. Nucl. Med,* **11:** 723–730, 1970.
5. P. J. Hurley, H. W. Strauss, and H. N. Wagner: Radionuclide Angiography and Cine-Angiography in Screening Patients for Cardiac Disease, *J. Nucl. Med.* **11:** 633, 1970.
6. T. P. Graham, J. K. Goodrich, A. E. Robinson, and C. Harris: Scintiangiocardiography in Children, *Am. J. Cardiol.* **25:** 387–394, 1974.

References

7. H. Wesselhoeft, P. J. Hurley, and H. N. Wagner: Nuclear Angiocardiography in the Differential Diagnosis of Congenital Heart Disease in Infants, *J. Nucl. Med* **12**: 406, 1971.
8. W. H. Oldendorf: Cerebral Blood Flow Measurements in *Recent Advances in Nuclear Medicine,* M. N. Croll and L. W. Brady, Eds., Appleton-Century Crofts, New York, 1966, pp. 44–70.
9. J. P. Kriss, L. P. Enright, W. G. Hayden, L. Wexler, and N. E. Shumway, Radioisotopic Angiocardiography: Findings in Congenital Heart Disease, *J. Nucl. Med.* **13**: 31–40, January 1972.
10. W. L. Ashburn, J. C. Harbert, W. C. Whitehouse, and D. T. Mason: A Video System for Recording Dynamic Radioisotope Studies with the Anger Scintillation Camera, *J. Nucl. Med.* **9**: 554–561, 1968.
11. W. E. Adam, P. Schenck, H. Kampmann, W. J. Lorenz, W. G. Schneider, W. Ammann, and L. Bilaniuk: Investigation of Cardiac Dynamics Using Scintillation Camera and Computer, Medical Radioisotope Scintigraphy, *Proc. Symp. Salzburg, 1968* **2**: 77, 1969.
12. M. A. Bender, L. Moussa-Mahmoud, and M. Blau, Quantitative Radio Cardiography with the Digital Autofluoroscope, Medical Radioisotope Scintigraphy, *Proc. Symp. Salzburg, 1968* **2**: 57, 1969.
13. Y. Ishii, and W. J. Macintyre: Measurement of Heart Chamber Volumes by Analysis of Dilution Curves Simultaneously Recorded by Scintillation Camera, *Circulation* **44**: 37–46, 1971.
14. W. J. MacIntyre, R. E. Botti, Y. Ishii, and W. H. Pritchard: Atraumatic Measurement of the Distribution of Myocardial Blood Flow Using ^{43}K and a Scintillation Camera, in *Medical Radioisotopes Scintigraphy 1972,* Vol II, International Atomic Energy Agency, Vienna, 1973.
15. R. H. Jones, D. C. Sabiston, B. B. Bates, J. J. Morris, P. A. W. Anderson, and J. K. Goodrich: Quantitative Radionuclide Angiocardiography for Determination of Chamber to Chamber Cardiac Transit Times, *Am. J Cardiol,* **30**: 855–864, 1972.
16. R. H. Jones, B. B. Bates, J. K. Goodrich, and C. C. Harris: Basic Considerations in Computer Use for Dynamic Quantitative Radionuclide Studies, Proceedings of Second Symposium on Sharing of Computer Programs and Technology in Nuclear Medicine, April 1972, CONF-720430, pp. 133–149.

PRINCIPAL EDITOR: ROBERT H. JONES

CONTRIBUTING EDITOR: PAGE A. W. ANDERSON

3. Congenital Heart Disease: Imaging and Analytic Methods

Evaluation of the patient with congenital heart disease involves identification of anatomic abnormalities and delineation of their hemodynamic significance. Although cardiac disorders are usually definitively assessed by cardiac catheterization, this diagnostic procedure involves a significant risk in the infant or severely ill patient. Development of reliable, noninvasive techniques able to provide the same information with less patient risk and discomfort would be an important advance in the management of cardiac disorders. Radionuclides are ideally suited to provide such noninvasive characterization of blood flow. The first attempt to evaluate congenital heart disease using radioactive tracers was reported in 1949 (1). A single unshielded Geiger–Muller tube was used to monitor radioactivity over the point of maximum cardiac impulse following intravenous injection of ^{24}Na. Patients with large left-to-right shunts demonstrated an early recirculation pattern clearly different from that in normal subjects. These observations emphasized the potential use of isotopes for evaluation of patients with congenital heart disease. However, neither the instrument nor the isotope were practical for general use, and clinical application of these techniques was not immediately realized. The development of the gamma-camera has renewed interest in radionuclide studies of congenital heart disease, and recent achievements in this area indicate substantial promise for the evaluation and serial followup of patients with intracardiac shunts.

General Considerations 33

This chapter summarizes the variety of radionuclide techniques used to assess congenital heart disease. Much of the published work in this area consists of isolated reports by individual laboratories, and further investigation will be required to delineate the eventual clinical applicabiltiy of these methods. Radionuclide cardiac studies do not provide the same information obtainable from cardiac catheterization. Specifically, pressure measurements and full anatomic localization of landmarks are neither now possible nor likely to come within the realm of noninvasive radioisotope measurements. However, these noninvasive techniques provide data concerning shunt pathways and quantitative flow measurements with less risk than cardiac catheterization. Quantitative nuclear cardiography may be used as a screening technique and has the advantage of being well suited for serial determinations. Future refinements in instrumentation and in techniques for radionuclide heart studies should enhance the general applicability of these studies.

GENERAL CONSIDERATIONS

Many radionuclide dynamic studies represent applications of the indicator-dilution technique, and observations obtained with radionuclides yield time–activity curves which are similar to those obtained with dye indicators (2, 8). Precordial scintigraphy permits counts to be related to the anatomic configuration of the heart so that both pictorial and quantitative data may be obtained. Anatomic abnormalities may be detected by noting variation from normal in the flow of a radionuclide bolus through the central circulation. Characteristic patterns of congenital disease may be recognized as they present patterns that can be correlated with findings on cineangiocardiography. Abnormal patterns of flow may also reflect hemodynamic alterations. Patients with a right-to-left intracardiac shunt can be recognized by an abnormally early entry of tracer into the left heart and systemic circulation, a decrease in the total amount of tracer introduced into the lungs during the first bolus passage, and a prolongation of tracer clearance from the central circulation due to abnormally early systemic recirculation of indicator. Three characteristics are typical of patients with left-to-right shunts: (*1*) Increased pulmonary blood flow associated with the shunt causes a rapid pulmonary transit of tracer which may be identified by a brief pulmonary transit time or by an abnormally early appearance of tracer in the left heart. However, measurable delay is required for pulmonary transit of tracer in left-to-right shunting, and left heart appearance of tracer is therefore not as immediate as with right-to-left shunting. (*2*) Early reappearance of tracer within the right-heart chambers distal to the site of the shunt. (*3*) Central recirculation of blood that prolongs tracer clearance from all cardiac chambers distal to the site of the shunt.

Each of the abnormalities associated with shunting is quantitatively related to the magnitude of the shunt. Data generated by an idealized mathematical model demonstrate the increasing distortion of curves associated with shunts of different magnitude (Fig. 18). With increasing shunt size, the time required for pulmonary transit is progressively less and the quantity of tracer recirculated through chambers distal to the shunt is progressively greater. Techniques for quantitation of shunts are based on the relationship of the magnitude of curve abnormality to the degree of

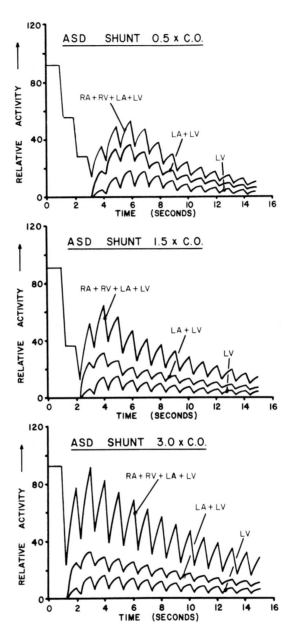

Fig. 18. Computer-constructed curves depicting expected alterations associated with left-to-right shunts of different magnitudes.

hemodynamic alteration. The concept of shunt quantitation by indicator-dilution techniques was introduced by Wood (3), and radionuclide techniques for shunt quantitation are based on this approach.

All limitations of indicator techniques for hemodynamic measurements exist also in radionuclide studies. Only the site of the most proximal shunt can be identified, and complex cardiac anomalies associated with more than one site of shunting may not be diagnosed accurately by radionuclide studies. In addition, severe abnormalities of tracer flow may present a confusing picture in unusual disorders. For example, greatly abnormal curve configurations associated with large shunts make difficult the separation of curve components necessary for accurate shunt calculation. Bidirectional shunting presents an additional source of difficulty, and indicator methods of shunt calculation in these situations are not reliable. Other potential inaccuracies of the technique result from limitations of external detection of radioactivity and the contamination of data with counts from several adjacent anatomic regions. For example, patients with coarctation of the aorta may appear to have a left-to-right shunt because of prolonged tracer clearance from the region of the heart, as activity appears in the bronchial arteries and therefore within the field of view, by collateral blood flow. The small heart size in children makes accurate anatomic selection of areas especially difficult. Small errors in delineation of borders of cardiac chambers will alter greatly the interpretation of data generated in a dynamic study.

The radionuclide most commonly used to assess congenital heart disease is ^{99m}Tc. It is generally accepted that the whole body dose to a 70-kg subject receiving ^{99m}Tc-pertechnetate intravenously is approximately 13 mrad/mCi (4,5). Gonadal dose is approximately the same in the male, but is slightly higher (15 mrad/mCi) in the female. When given as ^{99m}Tc-albumin the whole body dose is somewhat increased since there is no appreciable early excretion of the radiopharmaceutical. Based on a dose rate of 1.9 mrad/hr evenly distributed in the body, a whole body dose of 16 mrad/mCi is estimated (6). The extrapolation of these calculations for estimation of whole body and gonadal dose to children is not quite linear with mass since the absorbed fraction of the gamma-ray dose decreases concomitantly with volume.

Wellman, Kereiakes, and Branson (6) have estimated the whole body doses from ^{99m}Tc to children of various ages as shown in Table 1 (page 6).

A dose of 10 to 20 mCi for adults is usually given, and is held to provide benefits commensurate with the risk in terms of radiation exposure, particularly when compared with the alternative, diagnostic cinefluorography, which is associated with exposures between 4 and 100 rads/minute to the skin, and an exit dose of 0.3–12 rads/minute (7, 27).

SINGLE-PROBE TECHNIQUES

The initial precordial isotope dilution curves were obtained using a single probe placed over the apex of the heart and were successful in demonstrating differences between the time–activity curves of normal subjects and those patients with large left-to-right intracardiac shunts. The curve patterns recorded in patients whose shunt flow comprised greater than 30% of the pulmonary blood flow were consistently distinguishable from those in normal subjects by a sudden slowing in the decline of tracer activity after the initial abrupt fall from peak activity, which was followed by a prolonged time in clearing the tracer from the central circulation (12). Placement of the scintillation counter over the apex of either lung yields curves easier to interpret since there is no superimposition of dilution curves from the cardiac chambers. Analysis of the lung curves permits diagnosis but no localization of the site of the left-to-right shunt (9–13). More recent single-probe studies have been utilizing cardiac catheterization to select the site of central injection while monitoring over the right lung, heart, and head (14).

The advantage of radioisotopes over indocyanine green is that the same information can be obtained without using multiple catheters, making multiple injections, or withdrawing substantial blood samples, which are drawbacks of special concern in the infant or severely ill patient. The patterns observed appear the same in normals of all ages, including the infant, and changes in the curve configuration can be used to indicate early central recirculation of indicator.

Initially, a left-to-right shunt is recognized by recording the right lung and right heart curves following injection into the main stem or proximal right pulmonary artery. Early pulmonary recirculation of tracer with a left-to-right shunt results in a high continued activity over the heart and lung following the initial transit of the indicator through the lung and the heart. The site of the left-to-right shunt could then be determined by using selective injections at serial upstream sites in the left heart, while monitoring activity over the heart. The site of the distal shunt is localized by injecting sequentially into the ascending aorta, left ventricle, and left atrium. When the site of the shunt is reached, changes in the precordial radioactivity curve demonstrate the characteristics found in a left-to-right shunt.

In a patient with a ventricular septal defect, normal lung and heart curves are recorded following injection into the ascending aortic arch. However, following injection into the left ventricle, the right-lung activity increases to a level higher than that recorded following aortic root injection, and the heart curve shows the distortion of the disappearance phase due to early pulmonary recirculation (Fig. 19).

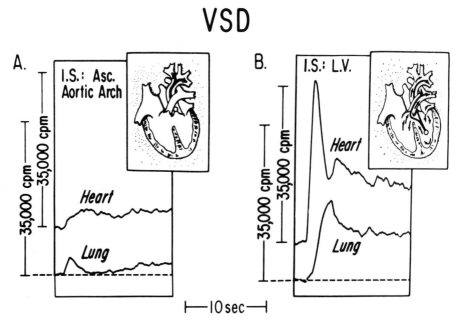

Fig. 19. Ventricular septal defect. (*A*) Normal curves followed ascending aortic arch injection. (*B*) Early and sustained increased lung activity with distortion of the heart curve was noted following left ventricular injection.

PATENT DUCTUS ARTERIOSUS

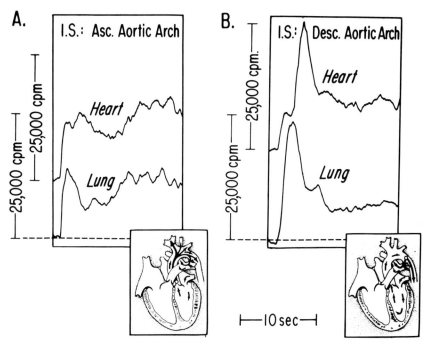

Fig. 20. *Patent ductus arteriosus.* (*A*) Injection near aortic valve produced a moderate early increase in lung activity. (*B*) Aortic injection near the ductus produced a more prominent increase in lung activity. Following injection near the ductus, a larger portion of the bolus of indicator was shunted across the ductus than occurred following aortic root injection.

The patient with the patent ductus arteriosus shows qualitatively quite different curves following an injection into the ascending aorta (Fig. 20). Injection into the ascending aorta near the aortic valve increases lung radioactivity immediately following the injection. After this peak lung activity, the heart curve demonstrates a count increase prior to the normal time of systemic recirculation. These abnormal findings are accentuated by injection into the aortic arch at the site of the ductus. On the other hand, a patient with an atrial septal defect shows normal heart and lung curves following aortic root and left ventricular injection. In this patient, left atrial injection produces a prominent rise in the right-lung curve that was not present after left ventricular injection (Fig. 21). A patient with partial anomalous venous drainage represents an extreme example of shunt localization. In this patient with drainage of the pulmonary veins of the right lung into the superior vena cava, injection into the distal left pulmonary artery yielded near normal left lung and heart curves, whereas injection into the right pulmonary artery produced lung and heart curves typical of a left-to-right shunt (Fig. 22). Coarctation of the aorta is one condition that will give apparent false positive left-to-right shunt heart curves following pulmonary artery and heart injections (Fig. 23). This curve abnormality results from collateral circulation through the internal mammary and intercostal arteries which are viewed by the precordial and lung probes.

ASD

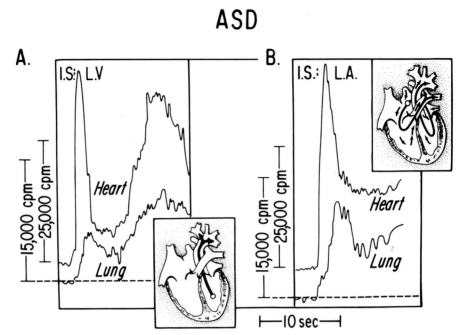

Fig. 21. Atrial septal defect (*A*) Left ventricular injection produced normal heart and lung curves except for the prominent recirculation rise. (*B*) Left atrial injection indicated a left-to-right shunt.

PARTIAL ANOM. PUL. VENOUS DRAINAGE

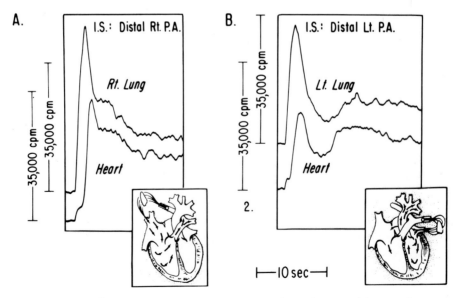

Fig. 22. Partial anomalous pulmonary venous drainage into the right superior vena cava. (*A*) Injection into the right pulmonary artery indicated a large left-to-right shunt. (*B*) Left pulmonary injection demonstrated near normal curves.

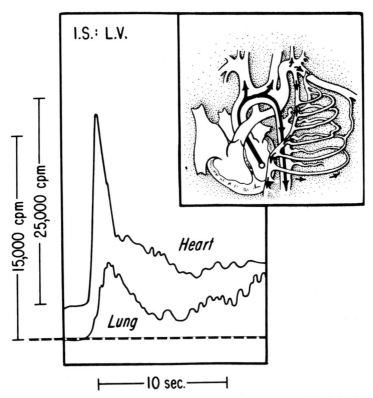

Fig. 23. Coarctation of the aorta. Following left ventricular injection, the precordial curve was distorted. Cine studies ruled out associated defects.

Flaherty et al. used radionuclide data from a single precordial probe to quantitate the magnitude of left-to-right shunts (20, 21). Studies were obtained at cardiac catheterization utilizing a tracer injection into the right pulmonary artery with counts recorded over the right lung to avoid background activity from cardiac chambers. The method of data analysis was similar to that described in detail in the following section on camera quantitation techniques. In 110 patients reported, the correlation coefficient was .94 between the Fick and the radionuclide techniques.

RIGHT-TO-LEFT SHUNTS

Single-proble studies have been of diagnostic value in patients with right-to-left shunts (22). Simultaneous acquisition of head and heart curves after injection first into the right and then the left heart provides the necessary data for demonstrating the presence of a right-to-left shunt. Following an injection into the right heart, the time required for appearance of tracer in the systemic circulation is apparent from an increase in counts over the head. If this time is similar to or shorter than that after the injection of tracer in the left heart, a right-to-left shunt is present, as tracer appears inappropriately early in the systemic circulation following a right-heart injection.

In a patient with severe pulmonary stenosis and an intact septum, injection into the right ventricle demonstrates a normal head and heart curve (Fig. 24). Injection into the right atrium demonstrates a right-to-left shunt by an early rise in radioactivity over the head at a time when tracer remains within the right heart.

Patients with total anomalous pulmonary venous drainage provide a characteristic combination of curves (Fig. 25). Following right ventricular injection a large left-to-right shunt is indicated by a sustained elevation in the precordial counts. A slight and gradual increase in the activity occurs over the head, beginning during the levophase of the precordial curve. However a right atrial injection in these patients demonstrates a large right-to-left atrial shunt by the early appearance of radioactivity over the head.

Transposition of the great vessels, which is the most common cause of cyanotic heart disease in the infant, provides a characteristic combination of curves (Fig. 26). Right ventricular injection produces a rapid rise in counts over the head, and

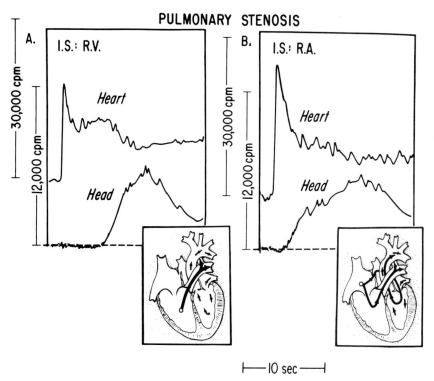

Fig. 24. Severe valvular pulmonic stenosis. (*A*) Normal heart and head curves were recorded following right ventricular (RV) injection. The rise in the head curve occurred at a time near peak activity of the left-heart portion of the precordial curve. (*B*) Right atrial (RA) injection demonstrated the right-to-left shunt by the rise in the head curve during inscription of the "right heart" portion of the precordial curve. The inserts illustrate the course of the circulation from the site of injection. The solid lines indicate initial traversal through the heart, and the interrupted arrows indicate return of the indicator from the lungs.

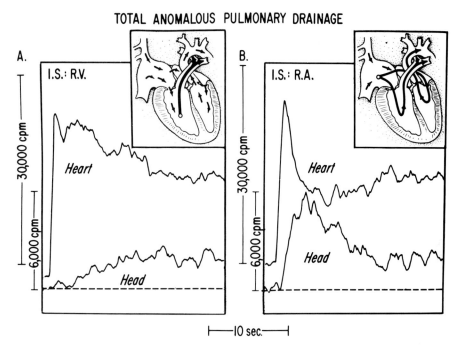

Fig. 25. Total anomalous pulmonary venous drainage. (*A*) The precordial curve indicates a large left-to-right shunt following right ventricular injection. The head curve indicates no right-to-left shunt. (*B*) The early increase in activity over the head after right atrial injection indicates the large right-to-left shunt at the atrial level.

the precordial curve demonstrates only a single peak with a rapid fall in radioactivity. Following left ventricular or left atrial injection, a prolonged appearance of increased precordial activity is associated with early, sustained pulmonary recirculation and a very slow increase in counts over the head. Patients with other cardiac anomalies that are physiologically similar to transposition of the great vessels demonstrate similar curves (23). An example of such an anomaly is double-outlet right ventricle with flow through the ventricular septal defect being committed to the pulmonary artery.

Single-probe studies during cardiac catheterization have proved useful for routine evaluation of patients with congenital heart disease. In the Duke Medical Center Laboratory, catheterization of each cyanotic infant is begun with radionuclide injections into the left atrium and right atrium or ventricle with recording over the heart and head. The resulting information aids selection of an orderly approach to the remainder of the catheterization.

CAMERA STUDIES

Visualization of Hemodynamic Alterations

Recently, studies have been reported using a gamma-camera for imaging the passage of a radionuclide bolus through the heart (15–19). Scintigraphs demonstrate characteristic abnormalities in a variety of congenital heart diseases.

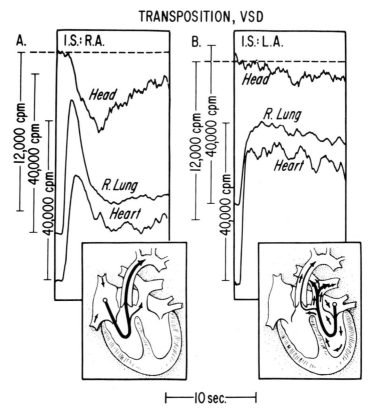

Fig. 26. Transposition of the great vessels in an 8-week-old infant with a small atrial defect and a ventricular defect with complete transposition of the great vessels. (*A*) Right ventricular (RV) injection. Following right ventricular injection there was a rapid fall to low-level activity in the precordial curve associated with a rapid rise in activity over the head. (*B*) Left ventricular (LV) injection. The precordial curve indicates continued pulmonary recirculation with slight increase in activity over the head.

Left-to-Right Shunts

Patients with an atrial septal defect demonstrate an enlarged right atrium evident on initial passage of radioisotope through the heart (Fig. 27a and 27b). The concentration of radioactivity decreases as blood passes from right atrium to right ventricle because of tracer dilution by unlabeled blood flowing through the atrial septal defect. The right ventricle and main pulmonary artery may also appear enlarged in the case of high-volume shunts. After the initial passage of the labeled blood through the lungs, early recirculation occurs with passage of radioactivity from the left-to-right atrium, with subsequent persistence of radioactivity in all cardiac chambers.

Patients with ventricular septal defects typically show no right atrial enlargement or any right atrial visualization after the first bolus passage through the lungs (Fig. 28). Instead, the right ventricle shows early appearance of activity after the return

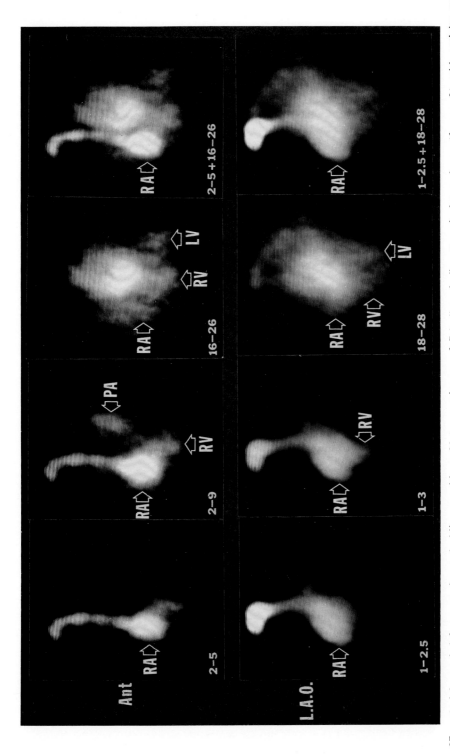

Fig. 27a. Atrial septal defect, anterior and oblique positions. Note prominence of RA, "smudge" pattern in later phases (frames 3) with activity present in RA (proved by double exposures, frames 4) and the other three heart chambers, failure to isolate an LV-aortic pattern, and prolonged visualization of intracardiac activity. Pulmonary/systemic flow ratio was 1.6.

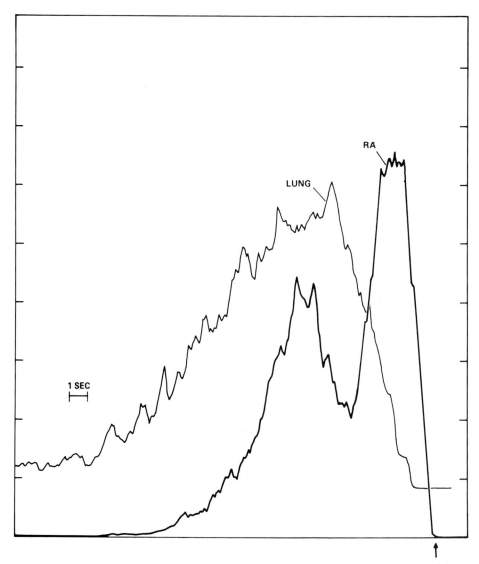

Fig. 27b. Time–activity curves (anterior view) for RA show a recirculation peak in the RA, and abnormally slow decline in lung activity, also due to continuous recirculation.

of labeled blood from the lungs. The lungs, the left atrium, and the right and left ventricles show persistence of radioactivity. Both atrial and ventricular septal defects cause poor delineation of the aorta because of proximal dilution of the tracer bolus.

Patients with aorticopulmonary communications, such as patent ductus arteriosus and truncus arteriosus, show a typical pattern in the absence of aortic insufficiency (Figs. 29a and 29b). Neither the right atrium nor the right ventricle show early reappearance of radioactivity, but as blood passes through the aorta both lungs demonstrate early reappearance of tracer. Subsequently, a high level of radioactivity continues over the left heart and the lungs.

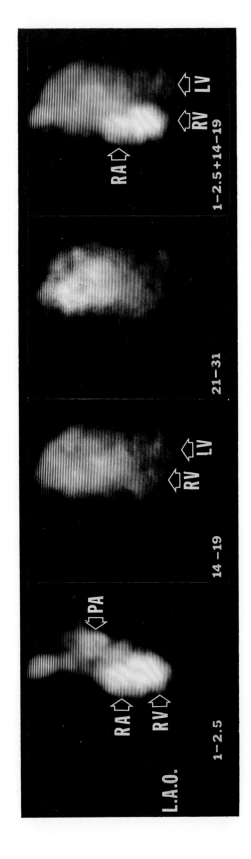

Fig. 28. Ventricular septal defect, oblique position. Note smudge patterns in later phases (frames 2 and 3), late visualization of RV and LV, but not RA (compare frames 2 and 4), and prolonged visualization of cardiopulmonary/systemic flow ratio was 1.2:1.

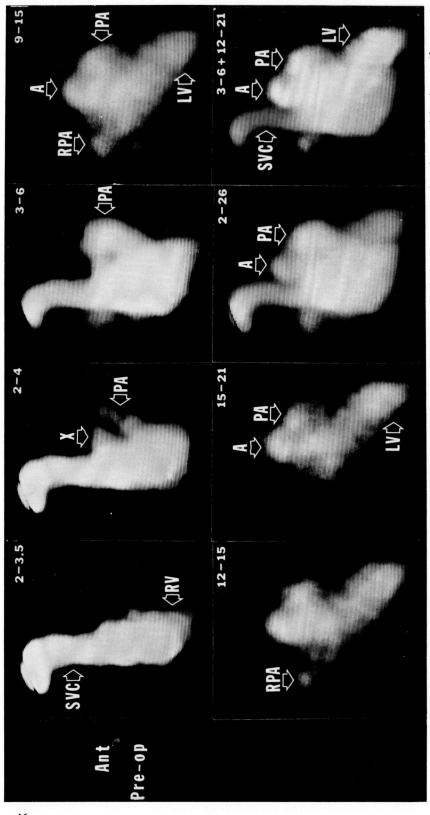

Fig. 29a. Patent ductus arteriosus, anterior view, pre- and postoperatively (6 days). Preoperatively, note somewhat larger right heart (frames 1 and 2) and early visualization of PA (frame 2), but persistent visualization of PA during filling of aorta (frames 3 and 4) without concomitant activity in RA or RV (frames 3 and 4). Postoperatively PA is well visualized early (frame 2), but is not seen during aortic filling (frames 5 and 6). Composites (frames 7 and 8) show smaller heart postoperatively, especially the LV.

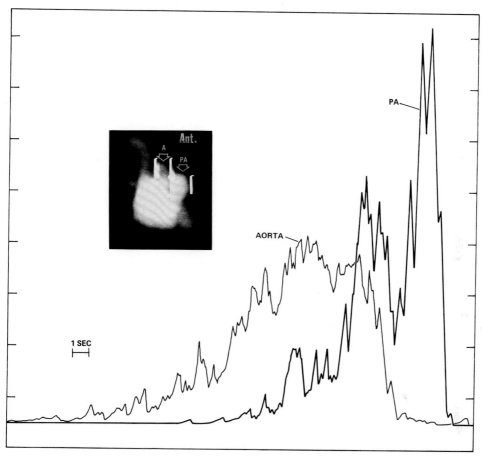

Fig. 29b. Analog activity curves (preoperative, anterior) with region-of-interest markers over aorta and pulmonary artery show double activity peak over PA and prolongation of aortic washout curve, typical of left-to-right shunt.

RIGHT TO LEFT SHUNTS

Demonstration of the presence of a right-to-left shunt without cardiac catheterization has been achieved by application of the scintillation camera (18) following right-side injection of tracer. Scintigraphs obtained in patients with right-to-left shunts typically demonstrate activity in the left heart and systemic vasculature before the lungs reach maximum tracer concentration (Fig. 30). In some patients, rapid sequence studies permit visualization of the location of the right-to-left shunt (Fig. 31). One limitation of this technique is that an enlarged right ventricle may be mistaken for the left ventricle. The major finding in the scintigraph which substantiates a right-to-left shunt is the earlier appearance of radioactivity in the systemic vasculature as compared with the pulmonic vasculature. An additional finding which substantiates the diagnosis of a large right-to-left shunt is poor delineation of the main pulmonary artery segment due to the passage of a substantial fraction of the tracer directly into the systemic circulation, bypassing the lung.

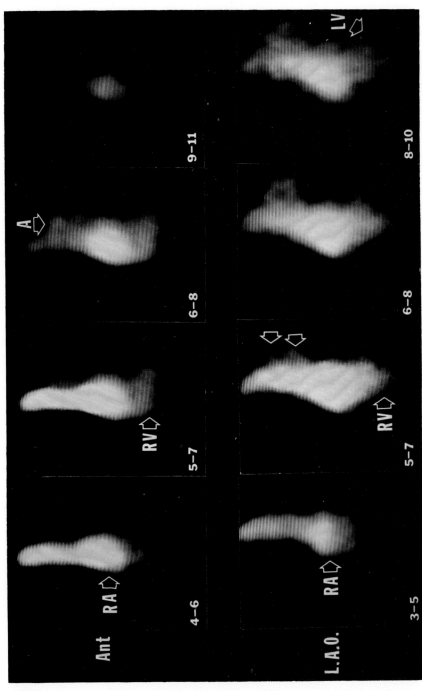

Fig. 30. Transposition of great vessels, anterior and oblique positions. Anteriorly note very early filling of aorta (A) (frames 2 and 3), rapid disappearance of cardiac activity (frame 4), failure to visualize left ventricle or pulmonary artery. In oblique position note filling of two outflow tracts after RV filling (arrows, frame 2, frame 3) corresponding to aorta and pulmonary artery joined by Blalock anastomosis (frame 3). LV is only faintly visualized (frame 4).

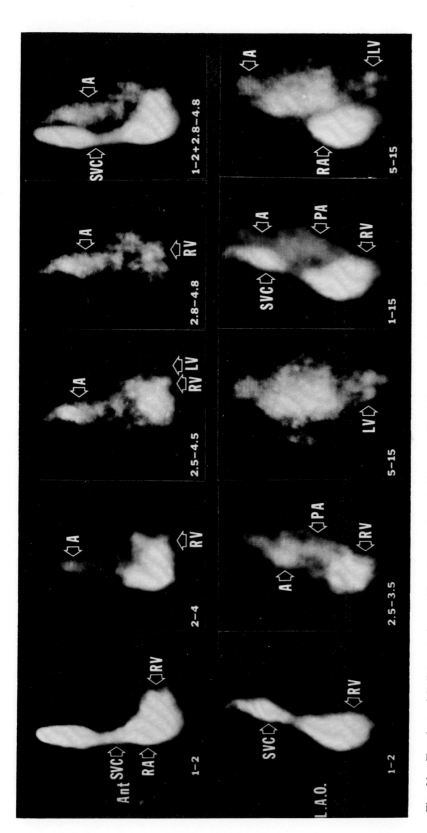

Fig. 31. Tetralogy of Fallot, anterior position (top row) and oblique position (bottom row). In anterior note very early filling of aorta (A) (frames 2–4) directly from RV (frame 4), but no visualization of pulmonic outflow tract. LV also was faintly visualized early in original scintiphoto (frame 3). In oblique study both PA and A are visualized early (frame 2), and LV was not seen until later (frame 3). Intense activity above RV (frame 3) was due to abnormal bronchial arterial channels, demonstrated more definitively by roentgenographic studies.

49

LEFT-TO-RIGHT SHUNT QUANTITATION

Radionuclide angiocardiography provides a means for quantitation of left-to-right cardiac shunts by venous tracer injection. Angiocardiographic data from the anatomic regions of the right atrium, right ventricle, main pulmonary artery, right lung, left atrium, left ventricle, and aortic arch demonstrate the characteristic abnormalities found in patients with left-to-right shunts (Fig. 32). Early recirculation to the right heart occurs at the level of the right ventricle in this patient with a ventricular septal defect.

The break in the normal curve decline in chambers distal to the shunt is due to indicator recirculation through the left-to-right shunt. This bypassing of the systemic circulation causes an early reappearance of labeled blood which is most apparent in the pulmonary circulation. A relatively constant portion of the labeled blood recirculates with each passage through the heart and the resulting series of early recirculation peaks causes an abnormal indicator persistence in the central circulation. The magnitude of curve distortion is directly related to the size of the shunt and provides an approach for shunt quantitation (21).

The accuracy of radionuclide measurement of left-to-right shunts was evaluated in 22 patients undergoing cardiac catheterization (28). Lung counts from radionuclide angiocardiograms are plotted on semilogarithmic coordinates. The initial descent of the exponential curve obtained is extrapolated to a value equivalent to 1% of the peak height of the curve (Fig. 33). The extrapolated line forms the division between two adjacent areas, x and y. The other borders of area y are the curve base line and a line extending from the curve peak perpendicular to the base

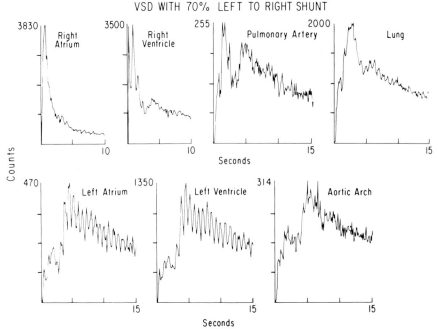

Fig. 32. Radionuclide data from a patient with ventricular septal defect and a 70% left-to-right shunt demonstrate typical findings in this disorder. Early appearance of tracer is apparent in the right ventricle curve indicating the shunt at this level.

Left-to-Right Shunt Quantitation

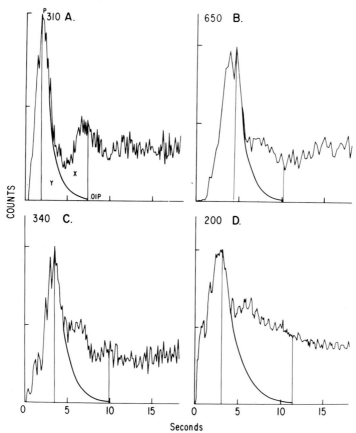

Fig. 33. Analysis of right-lung angiograms: (A) normal, (B) ventricular septal defect, 26% left-to-right shunt, (C) partial anomalous pulmonary venous drainage, 46% left-to-right shunt, (D) ventricular septal defect, 70% left-to-right shunt. The area ratio x to y as used for comparison with the calculated percent left-to-right shunt by the Fick method. Note the prominent break in the exponential decline in patients B, C, and D with the early central recirculation due to the shunt. (P) Peak value of the curve; $.01P$, 1% of the peak value.

line. The other borders of area x are the actual curve and a line perpendicular to the base line at the point where the extrapolation reaches 1% of the peak value. The final data analysis utilizes the ratio of these two areas, which obviates the need for actual scale factors for curve calibration. Areas are measured by planimetry, or by computer summing if a digital system is being used, and the ratio x/y is compared graphically and statistically to the magnitude of the shunt calculated from the Fick data.

In Figure 34 the data for the patients with left-to-right shunts and those with no cardiovascular abnormality are presented with the solution of the quadratic equation, which provided a correlation coefficient of .951 (standard deviation 6.8% and standard error 7.1%). When the size of the shunt exceeds 70% of pulmonary blood flow, the solution of the equation is stopped as it gives a paradoxical decrease in the size of the shunt as the ratio of the areas became larger. This limitation to

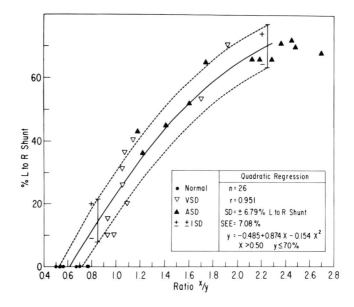

Fig. 34. Regression analysis of normals and patients with left-to-right shunts relating percent left-to-right shunt to the ratio x/y. N, Number of patients in regression analysis; r, correlation coefficient; SD, standard deviation; SEE, standard estimated error; ASD, atrial septal defect; and VSD, ventricular septal defect.

the method of analysis is of no clinical significance because the management of patients with 70% or larger left-to-right shunts is the same.

All normal patients had a ratio of x/y less than 0.8, while all patients with shunts in excess of 70% had an area in excess of 1.9. Between these values there was a monotonic increase in the ratio which correlated with the increase in the shunt flow from 30 to 70% of the pulmonary blood flow. Additionally, the method detected shunts sufficiently small to be seen by cineangiocardiography, but not by the Fick method. No false positive values appeared in the normal group.

This radionuclide method of estimating the size of left-to-right shunts has the limitations inherent in indicator-dilution techniques. Patients with severe mitral valvular or aortic valvular insufficiency demonstrate slurring of the downslope of the lung curve, producing an abnormally high x/y ratio (Fig. 35). Although the ratios suggest large left-to-right shunts in these patients, the prolonged pulmonary transit time is not consistent with a left-to-right shunt. In addition, patients with coarctation of the aorta may appear to have a left-to-right shunt due to intercostal artery flow. Also, the presence of a significant bidirectional shunt prevents accurate analysis of either component of shunting.

This study demonstrates that radionuclide angiograms can be used to quantitate left-to-right shunts with high accuracy when the ratio-of-areas method of analysis is used. The method permits detection of the small left-to-right shunt in patients with equivocal physical findings. In addition, patients with large shunts can be followed by serial assessment of shunt magnitude, thereby aiding the physician in deciding whether to catheterize when a decrease in shunt flow is associated with

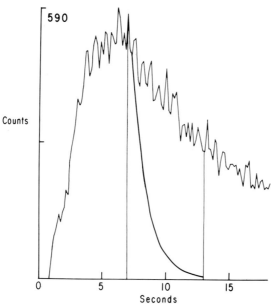

Fig. 35. Analysis of the right-lung angiogram of a patient with severe mitral insufficiency. Note the absence of a single exponential curve with a definite break point compared to curves in Figure 18 for patients with shunts. An attempt to analyze the curve provides an area ratio much greater than the mean ratio for the normals.

clinical findings suggesting the onset of Eisenmenger's syndrome. Moreover, the technique would appear promising in evaluating patients after surgical closure of septal defects, imposing little risk, discomfort, or radiation exposure to the patient, when compared with cineangiocardiography.

RIGHT-TO-LEFT SHUNT QUANTITATION

Quantitative radionuclide angiography is a potential approach for measuring the degree of a right-to-left shunt (17, 28). The anatomic resolution of this technique allows the isolation of cardiac chambers and the great vessels necessary for shunt localization. Early tracer appearance in the left heart and systemic circulation indicates right-to-left shunting. In a patient with tetralogy of Fallot, quantitative angiographic data demonstrate early tracer appearance in the left ventricle which is the site of the shunt (Fig. 36). Quantitation of right-to-left shunts by comparison of areas under indicator dilution curves would appear feasible from radionuclide angiogram data, although no report of such studies is now available.

Another technique which has been utilized to measure the magnitude of right-to-left shunts involves injection of radionuclide-tagged macroaggregates of serum albumin (MAA) (24–26). In the normal circulation, particles larger than capillary size are filtered in the pulmonary vasculature and do not reach the systemic circulation. In a patient with a right-to-left shunt, a fraction of intravenously injected

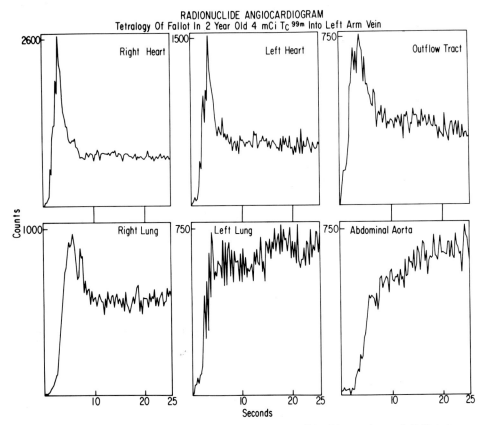

Fig. 36. Radionuclide angiogram data in a 2-year-old child with tetralogy of Fallot demonstrates characteristics typical of right-to-left intracardiac shunting. Tracer appears in the left heart and systemic circulation immediately after right-heart appearance.

particles bypasses the pulmonary circulation and lodges in other body organs. Lung counts may be compared with total body counts to calculate the magnitude of right-to-left shunt, if equipment available can provide quantitative measurements of radioactivity from the whole body for comparison with lung counts. However, the intravenous injection of MAA in children with cyanotic heart disease and right-to-left shunting of blood imposes the potential hazards of systemic embolism of the particles. MAA studies in more than 100 patients with right-to-left shunting have been reported with no indication of ill effects from systemic embolism. However, the potential remains for significant systemic embolism, and MAA lung scans in patients with right-to-left shunts must be approached with caution.

Figure 37a shows a visual presentation of an aortopulmonary window; Figure 37b shows an analog tracing of the transit characteristics of the bolus in the right atrium, left atrium, and aorta in this case.

Figure 38a shows the characteristics of tricuspid atresia, and Figure 38b shows the indicator dilution curve characteristics from right ventricle and right lung, showing the marked delay of transit through the enlarged right atrium.

Pseudotruncus is illustrated in Figure 39, showing the early pulmonary filling

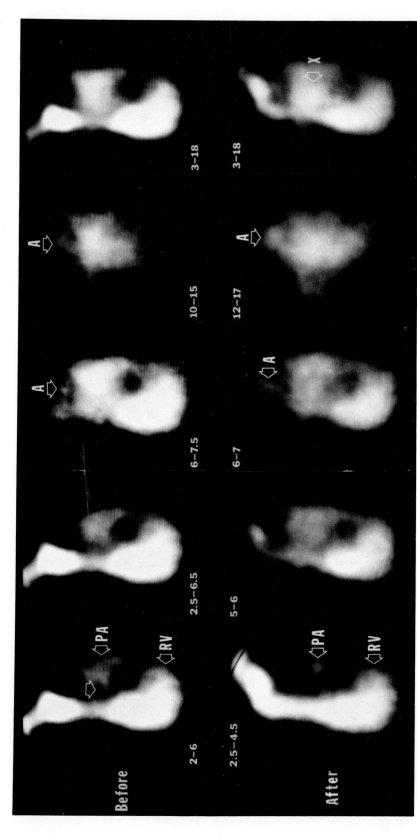

Fig. 37a. Aortopulmonary window, anterior position, before and after brief walking exercise. Note asymmetry of PA to the right, with tract filling ascending aortic area (frames 1 and 2, top arrow), early aortic (A) activity (frame 3), and relatively increased prominence of ascending aortic region after exercise (compare X, frame 5, before and after exercise).

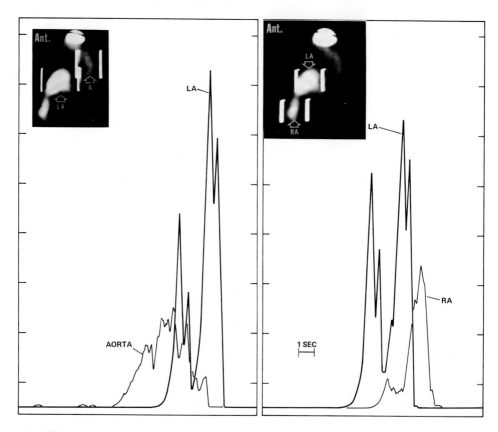

Fig. 37b.

following visualization of the aorta. Ebstein's anomaly, tricuspid atresia with downward displacement of the tricuspid valve, is shown in Figure 40, which indicates the very large atrium and delayed filling of the left heart.

Figures 41 and 42 show congenital and supra valvular aortic stenosis, respectively. Poststenotic dilatation of the aorta is characteristic of the former, and a visible narrowing or even nonvisualization of the aortic root is typical in the latter congenital form of aortic disease. Pre- and postoperative displays are shown.

Pulmonic stenosis is seen in Figure 43, with characteristic small right ventricle and poststenotic dilatation of the pulmonary artery. Clearing of tracer is slow from the large-volume pulmonary artery.

SUMMARY

Radionuclide studies can provide hemodynamic information which is useful in the diagnosis and management of patients with congenital heart disease. Studies using single detectors have provided the greatest information when combined with routine cardiac catheterization studies. Anatomic localization of the injection site aids localization of the site of abnormal blood flow. The use of a radioactive tracer obviates

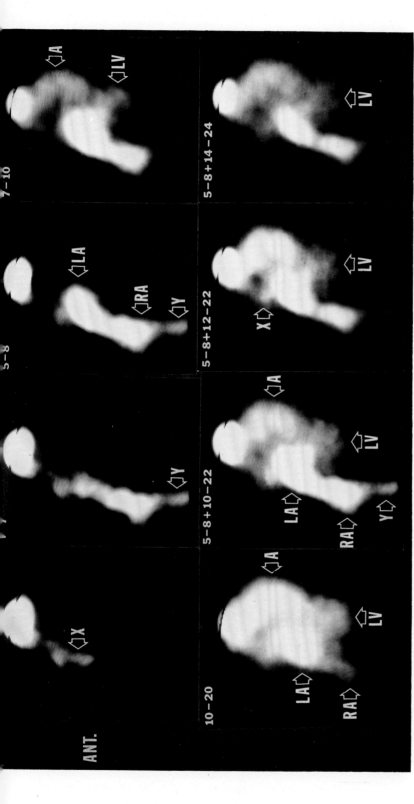

Fig. 38a. Tricuspid atresia, superior caval obstruction, anterior view. Probably due to previous surgery, venous bolus inflow was partially held up in upper mediastinum (frames 1–3) and a venous channel (X) coursed posterior to the heart (proved by a second study in RAO position) and filling of the RA (frame 3) occurred by the inferior vena cava (Y) (frames 2 and 3). Because of the tricuspid atresia, after traversing the RA, bolus transit was directed upward and to the left directly to the LA (frame 3), and thereafter to the LV and aorta (frame 4). Simultaneously with LV filling, a hypoplastic RV filled across a VSD (region between RA and LV, frame 5). Blood arriving later back into the LA via hypoplastic R–pulmonary artery–lung–LA circuit caused visible reentry peak over the LA as shown by the analog accompanying time–activity curves over selected regions of interest.

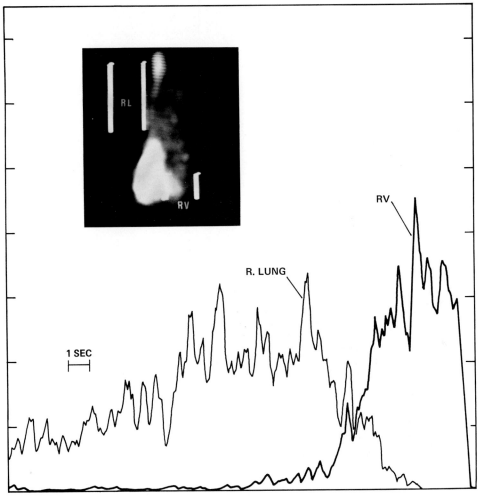

Fig. 38b.

need for arterial sampling, which is an especially important consideration in children. The spatial orientation of observed counts made possible by the gamma-camera permit both a pictorial and a quantitative description of the transit of a radionuclide bolus through the central circulation. Quantitative studies which include data from discrete cardiac regions provide an approach for measurement of the hemodynamic alterations observed. An especially promising area for quantitative studies is the localization and quantitation of left-to-right intracardiac shunts. Initial studies suggest that techniques of shunt quantitation using this approach are sufficiently accurate to be of use in initial screening and serial evaluation of patients with these disorders. The recognition and quantitation of right-to-left intracardiac shunts using radionuclide tracers represents a promising approach for evaluation of the cyanotic newborn. Continued improvements in instrumentation for detection of rapidly changing radioactivity promises to extend greatly the application of these studies for evaluation of cardiac anomalies and their associated hemodynamic alterations.

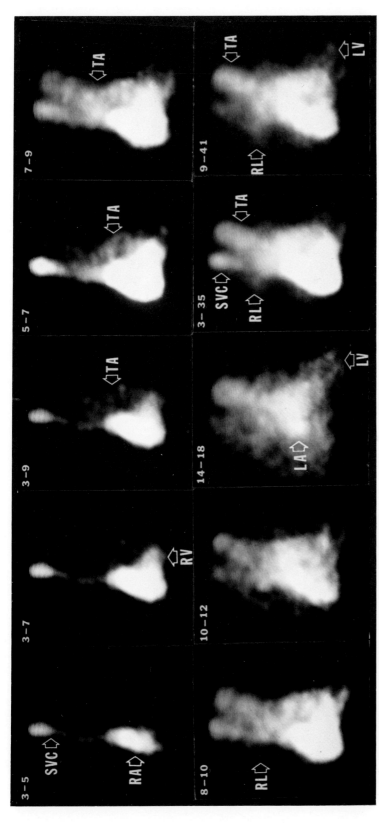

Fig. 39. Pseudotruncus arteriosus, anterior position. Note early filling of a large single vessel (truncus, TA) in frames 3–5, but no pulmonary artery and *no pulmonary activity until after 9 seconds* (frames 6 and 7), slightly delayed filling time of LA and LV (frame 8), and relatively high position of the arteries to the lung derived from the aorta (RL, frames 6, 9, and 10).

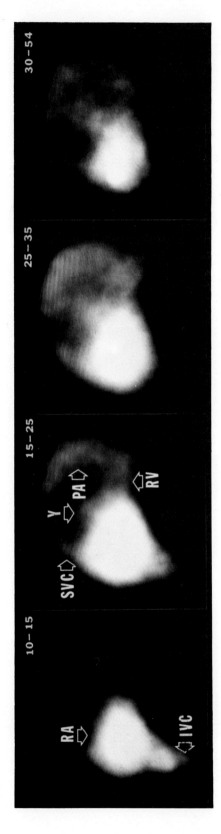

Fig. 40. Ebstein's anomaly, right anterior oblique position. Injection was made via an inferior vena cava (IVC) catheter. Note huge RA (all frames), reflux into SVC (frame 2), downward (lateralward) displacement of tricuspid valve with creation of a second "atrialized" chamber (Y) and markedly prolonged bolus transit.

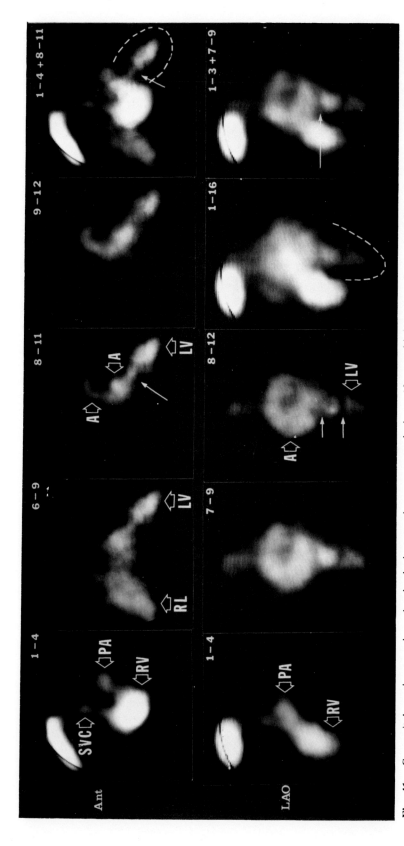

Fig. 41. Congenital aortic stenosis and subvalvular membranous stenosis in a 13-year-old boy, anterior and oblique positions. Anteriorly (top) note narrowing in region of aortic valve (arrow, frames 3 and 5), wide separation of LV and RV cavities (frame 5) and poststenotic dilatation of aorta (frame 3). In oblique position, poststenotic aortic dilatation is seen best in frames 2 and 3. In frame 3 top arrow shows position of aortic valve, bottom arrow probable position of ventricular membrane. Dotted lines indicate the estimated outer limits of LV myocardial wall; LV hypertrophy was present.

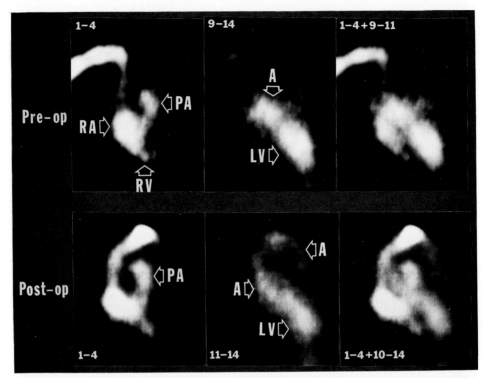

Fig. 42. Supravalvular aortic stenosis, anterior position, pre- and postoperatively (1 week). Preoperatively note sharp cut-off of activity just above aortic root (A) at point of stenosis and failure to visualize ascending aorta (frame 2), and LV dilation. Postoperatively note normal aortic filling (frame 2) and slightly smaller LV.

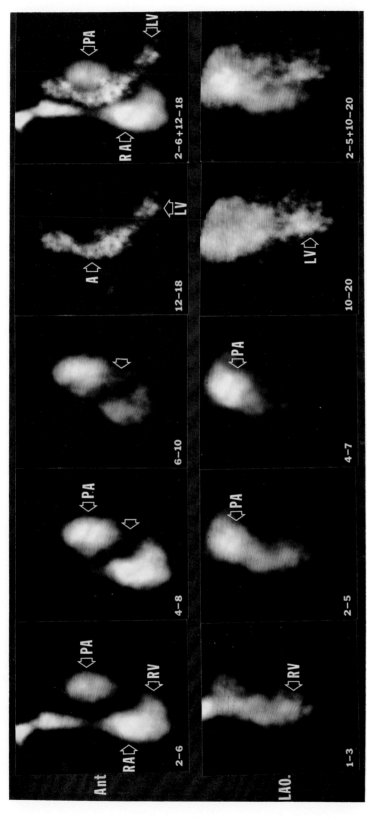

Fig. 43. Pulmonic stenosis, anterior and oblique positions. Anteriorly note relatively small-sized cavity of RV (frames 1 and 2), persistent narrowing in right ventricular outflow tract due to hypertrophied infundibulum (arrow, frames 2 and 3), persistent filling of dilated main pulmonary artery (PA) just beyond narrowing, and delayed filling of LV and aorta. In oblique position, note RV cavity is relatively small and PA is dilated and retains activity over long interval (frames 2 and 3).

REFERENCES

1. M. Prinzmetal, E. Corday, R. J. Spritzler, and W. Fleig: Radiocardiography and Its Clinical Application, *J. Am. Med. Assoc.* **139**: 617–622, 1949.
2. E. H. Wood, H. J. C. Swan, and H. F. Helmholtz: Recording and Basic Patterns of Dilution Curves; Normal and Abnormal. *Proc. Staff Meeting Mayo Clinic* **32**: 464, 1957.
3. E. H. Wood: Diagnostic Applications of Indicator-Dilution Techniques in Congenital Heart Disease, *Circulation Res.* **10**: 531–568, 1962.
4. E. M. Smith: Internal Dose Calculation for ^{99m}Tc, *J. Nucl.* **6**: 231–251, 1965.
5. R. M. Cloutier, and E. E. Watson: Medical Radionuclides: Radiation Dose and Effects, Atomic Energy Commission Symposium Series 20, Oak Ridge, Tennessee, 1970, pp. 325–346.
6. H. N. Wellman, J. G. Kereiakes, and B. Branson: ibid, pp. 133–155.
7. N. G. Trott, A. J. Stacey, R. E. Ellis, and F. M. Dermentzoglou: Ibid, pp. 157–183.
8. Y. Van Der Feer, J. H. Douma, and W. Klip: Cardiac Output Measurement by the Injection Method Without Actual Sampling, *Am. Heart J.* **56**: 642–651, 1958.
9. J. P. Turner, E. Falazar, and R. Gorlin: Detection of Intracardiac Shunts by an External Surface-Counting Technique, *Clin. Res.* **7**: 230, 1959.
10. R. L. Huff, D. Parrish, and W. Crockett: Study of Circulatory Dynamics by Means of Crystal Radiation Detectors on the Anterior Thoracic Wall, *Circulation Res.* **5**: 395–400, 1957.
11. R. H. Greenspan, R. G. Lester, J. F. Marvin, and K. Amplatz: Isotope Circulation Studies in Congenital Heart Disease, *J. Am. Med Assoc.* **169**: 667–672, 1959.
12. W. Shapiro, A. R. Sharpe: Precordial Isotope-Dilution Curves in Congenital Heart Disease, a Simple Method for the Detection of Intracardiac Shunts, *Am. Heart J.* **60**: 607–617, 1960.
13. R. Folse, E. Braunwald: Pulmonary Vascular Dilution Curves Recorded by External Detection in the Diagnosis of Left-to-Right Shunts, *Brit. Heart J.* **24**: 166–172, 1962.
14. M. S. Spach, R. V. Canent, J. P. Boineau, A. W. White, Jr., A. P. Sanders, and G. J. Baylin: Radioisotope-Dilution Curves as an Adjunct to Cardiac Catheterization: I. Left-to-Right Shunts, *Am. J. Cardiol.* **16**: 165–175, 1965.
15. H. O. Anger: Scintillation Camera, *Rev. Sci. Instr.* **29**: 27–33, 1958.
16. D. T. Mason, W. L. Ashburn, J. C. Harbert, L. S. Cohen and E. Braunwald: Rapid Sequential Visualization of the Heart and Great Vessels in Man Using the Wide Field Anger Scintillation Camera. *Circulation* **39**: 19–28, 1968.
17. T. P. Graham, J. K. Goodrich, A. E. Robinson, and C. C. Harris: Scintiangiocardiography in Children, *Am. J. Cardiol.* **25**: 387–394, 1970.
18. J. P. Kriss, L. P. Enright, W. G. Hayden, L. Wexler, and N. E. Shumway: Radioisotopic Angiocardiography: Wide Scope of Applicability in Diagnosis and Evaluation of Therapy in Diseases of the Heart and Great Vessels, *Circulation* **43**: 792–808, 1971.
19. R. H. Jones, D. C. Sabiston, B. B. Bates, J. J. Morris, P. A. W. Anderson, and J. K. Goodrich: Quantitative Radionuclide Angiocardiography for Determination of Chamber to Chamber Cardiac Transit Times, *Am. J. Cardiol.* **30**: 855–864, 1972.
20. J. T. Flaherty, R. V. Canent, J. P. Boineau, P. A. W. Anderson, A. R. Levin, M. S. Spach: Use of Externally Recorded Radioisotope Dilution Curves for Quantitation of Left-to-Right Shunts, *Am. J. Cardiol* **20**: 341–345, 1967.
21. N. P. Alazraki, W. C. Ashburn, A. Hogan, and W. F. Friedman: Detection of Left-to-Right Cardiac Shunts with the Scintillation Camera Pulmonary Dilution Curves, *J. Nucl. Med.* **13**: 142–147, 1972.
22. M. S. Spach, R. V. Canent, J. P. Boineau, et al: Radioisotope-Dilution Curves as an Adjunct to Cardiac Catheterization: II. Right-to-Left Shunts, *Am. J. Cardiol.* **16**: 176–183, 1965.

References

23. P. J. Hurley, H. W. Strauss, and H. N. Wagner, Jr.: Radionuclide Angiocardiography in Cyanotic Congenital Heart Disease, *Hopkins Med. J.* **127:** 46–54, 1970.
24. G. V. Taplin, J. C. Kennedy, M. S. Griswold, M. M. Akcoy, and D. E. Johnson: Albumin ^{125}I Macroaggregates for Brain Scanning. *J. Nucl. Med.* (Abst.) **5:** 366, 1964.
25. C. Y. Lin: Lung Scan in Cardiopulmonary Disease: I. Tetralogy of Fallot, *J. Thoracic Cardiovasc. Surg.* **61:** 370–379, 1971.
26. J. C. Kennedy, G. V. Taplin: Albumin Macroaggregates for Brain Scanning; Experimental Basis and Safety in Primates, *J. Nucl. Med.* **6:** 566–581, 1965.
27. R. L. Penfil, M. C. Brown: Genetically Significant Dose to the United States Population for Diagnostic Medical Roentgenology, 1964, *Radiology* **90:** 209–216, 1968.
28. P. A. W. Anderson, R. H. Jones and D. C. Sabiston, Jr.: Quantitation of Left-to-Right Cardiac Shunts with Radionuclide Angiography, *Circulation* **49:** 512–516, 1974.

PRINCIPAL EDITOR: JOSEPH P. KRISS

4. Acquired Cardiovascular Disease

Noninvasive dynamic radionuclide studies of the heart and great vessels permit the evaluation of numerous acquired and congenital lesions. A wide spectrum of acquired cardiovascular diseases has been studied (9), which includes the more commonly observed entities such as valvular heart disease (11), ventricular and aortic aneurysms (1, 4) as well as less commonly observed conditions such as superior caval obstruction (6), pericardial effusion (1, 2), cardiac tumors, sinus of Valsalva aneurysm (3, 10), cardiomyopathy, idiopathic pulmonary hypertension (5), hyperkinetic states, and luetic aortitis. In this chapter the diagnostic criteria employed for acquired lesions are presented together with illustrative examples to demonstrate the type and quality of the information that may be obtained with the method.

The reader is referred to Chapter 8 for description of the various instrumentation systems which may be employed (7), such as the variable time-lapse videoscintiscope (VTV) (8), digital computer–scintillation camera systems and autofluoroscope, and to Chapter 2 for a description of recommended positioning and injection techniques, and for descriptions of the scintigraphic and bolus transit findings which characterize the normal heart. The case examples which follow were studied using the VTV and/or a digital computer*–scintillation camera† system. Most patients have been studied in more than one position, usually anterior and modified left anterior oblique (LAO), less frequently right anterior oblique (RAO).

VALVULAR HEART DISEASE

Mitral Stenosis

Mitral stenosis without insufficiency is associated with the following features: (*1*) left atrial enlargement; (*2*) normal left ventricular size in the absence of other

* Hewlett-Packard, Model 5407A Scintigraphic Data Analyzer.
† Searle Radiographics, Pho-Gamma III.

underlying lesions; (*3*) prolonged visualization of the left atrium and left ventricle, with persistent relatively greater radioactivity in the left atrium (Fig. 44); (*4*) with greater degrees of stenosis, or in more long-standing cases, in addition to the above, enlargement of the pulmonary outflow tract, right ventricular enlargement, and prolongation of the cardiopulmonary circulation time which is a function of the severity and duration of the stenosis (Fig. 45).

Repeat and serial studies may be performed readily after mitral commisurotomy to follow the postsurgical progress of anatomic and hemodynamic changes (10) (Fig. 46). Even a few days following surgery one may detect evidence of improvement, such as shortened circulation time, decreased size of the left atrium, improved filling of the left ventricle and aorta, and overall decrease in heart size. However, complete regression to a normal-sized left atrium has not been demonstrated in the early postoperative period despite marked clinical improvement.

Mitral Insufficiency

Because the hemodynamic consequences of mitral regurgitation impose volume work on the left atrium and the left ventricle, as opposed to the pressure work of mitral stenosis, the characteristic features of this lesion differ and can be summarized as follows: (*1*) left atrial enlargement is usually more marked than in cases with stenotic lesions of the mitral valve; (*2*) left ventricular dilation which reflects the increased volume work of this chamber; (*3*) persistent visualization of radioactivity for 5 seconds or more in both the left atrium and the left ventricle, secondary to the to-and-fro movement of tracer across the incompetent valve (12). These features are shown in Figure 47, which also demonstrates that operative replacement of the mitral valve may be accompanied by striking and rapidly developing evidence of improvement, but that enlargement of the left atrium remains angiographically demonstrable for at least several weeks after surgery.

Left atrial enlargement, due to mitral stenosis and/or mitral insufficiency, may occasionally be demonstrated better in the right anterior oblique view than in other positions. This phenomenon is illustrated in Figure 48. In this patient the criteria for mitral insufficiency cited above are met in the LAO view, but the left atrial enlargement is better seen in the RAO position. The latter position is preferred by some investigators recording high-frequency changes in radioactivity within the left ventricle because that position permits optimal anatomic separation of LV from the LA and aorta. In the RAO position, the left ventricle is viewed through the intervening right ventricle.

In patients with extreme degrees of dilatation of the left atrium, an anterior study may show that, due to its posterior and rightward displacement, the left atrium presents above and to the right of the right atrium (Fig. 49).

An unusual delineation of the left atrium has been observed in several patients in association with left ventricular aneurysm, both before and after repair of the aneurysm (Figs. 50 and 51). In these instances, the left atrium has not appeared to be so much enlarged as deformed. It is postulated that the phenomenon observed might be related to valvular dysfunction related to disease of one or more papillary muscles.

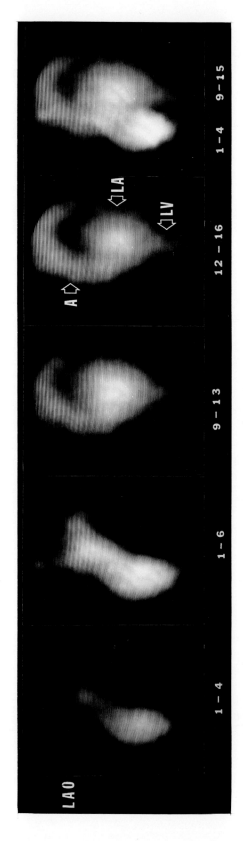

Fig. 44. Mitral stenosis, oblique position. Note large LA (frames 3–5) with prolonged residence time and small LV (frames 3–5). The aorta is unusually well seen in this case.

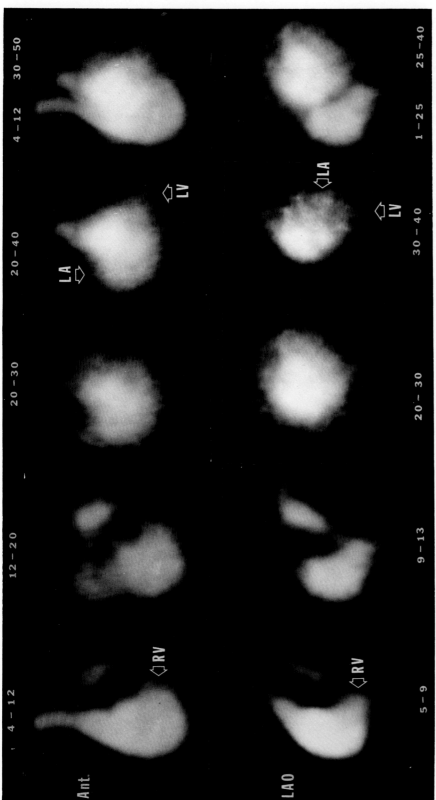

Fig. 45. Combined mitral stenosis and aortic stenosis, pulmonary hypertension and congestive heart failure. In anterior views note enlargement of RA and RV (frame 1), prolonged bolus residence in the right heart (frame 2), marked LA enlargement (frames 3 and 4), small LV cavity (frames 4 and 5), and widening of the IV septum (oblique, frame 5). A diagnosis of aortic stenosis, while consistent with the findings, could not be definitely made, since the widening of the IV septum could be due to hypertrophy of the RV wall caused by the pulmonary hypertension.

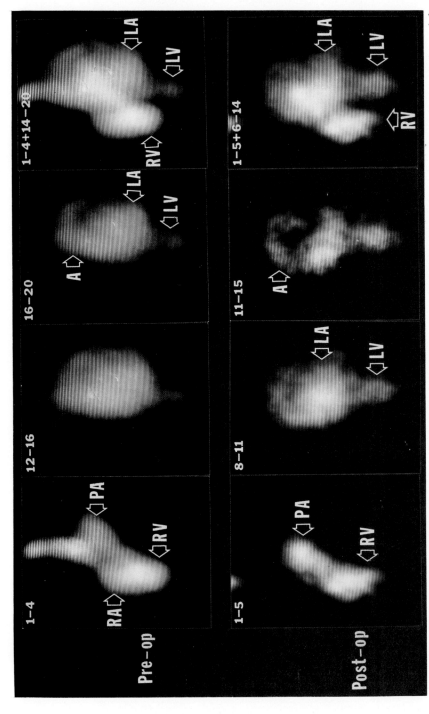

Fig. 46. Mitral stenosis. Oblique position, pre- and postoperatively (2 months). In the preoperative series note very large left atrium (frames 2–4), normal left ventricle, and prolonged residence of activity in left atrium with activity dominance in left atrium, compared to left ventricle. Two months after operation, note slightly smaller heart size (frame 4), shorter circulation time, better filling of left ventricle, and still enlarged but smaller left atrium with improved emptying of that chamber (frames 2 and 3). Little activity remained in the heart after 15 seconds (frames not shown).

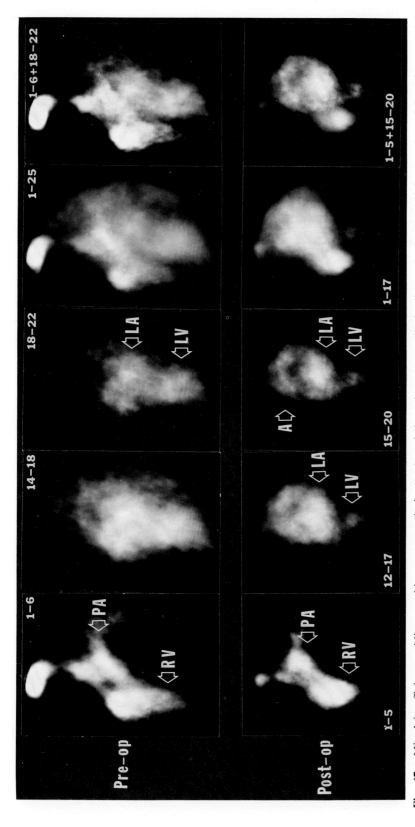

Fig. 47. Mitral insufficiency, oblique position, preoperatively (top) and 6 weeks postoperatively (bottom). Preoperatively note large left atrium and left ventricle (frames 2–5), poor delineation of aorta, and prolonged visualization of left-heart chambers with relatively equal intensity of activity in left ventricle and left atrium. Six weeks after operation, note smaller heart generally, enlarged but smaller left atrium, greatly reduced size of left ventricular cavity and normal aorta.

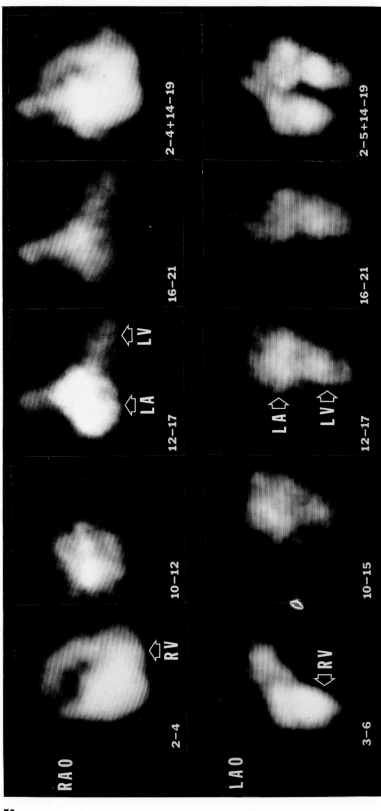

Fig. 48. Mitral insufficiency. Right anterior oblique, RAO, (top) and modified left anterior oblique, LAO (bottom) positions. In RAO study, RA and RV are prominent (frame 1), enlarged LA is well shown (frames 2–4) with prolonged residence there, and LV and aorta are thrown relatively clear of the LA (frames 3 and 4), but the composite view (frame 5) is not helpful in separating left- from right-sided cardiovascular structures. In the LAO study, enlarged LA is seen (frames 2–4); there is approximately equal activity in both LA and LV over a long time period (frames 3 and 4) and the composite view (frame 5) shows good delineation of the cardiac chambers.

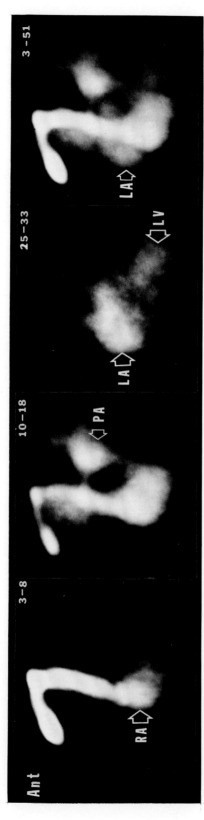

Fig. 49. Mitral stenosis and insufficiency, anterior view. Right atrium (RA) and right ventricle (RV) are enlarged, and pulmonary artery (PA) is very prominent (frame 2). Left atrium (LA) is greatly dilated and intensely delineated compared with left ventricle (LV), which is also enlarged (frame 3). Frame 4 demonstrates in striking fashion how left atrium has extended above and to the right of right atrium.

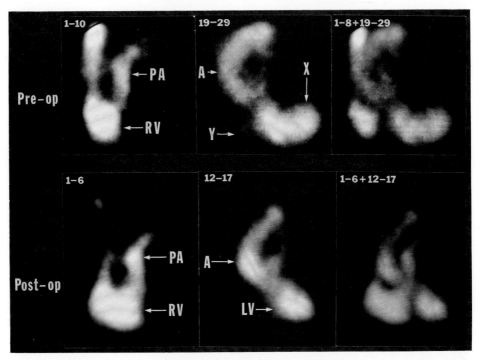

Fig. 50. Left ventricular aneurysms, anterior view, pre- and postoperative. Preoperatively (top row), note two irregular gross dilatations of the left ventricle with aneurysmal ballooning at sites X and Y (frames 2 and 3) and the prolonged circulation time. One year after removal of both aneurysms and ventricular patching, the left ventricle appears normal and circulation time is improved.

Aortic Stenosis

Pure aortic stenosis presents the left ventricle with an obstructive lesion that is initially compensated by increased pressure work in that chamber. (*1*) The left ventricular cavity is decreased in volume secondary to ventricular wall hypertrophy; this remains a valid finding up to the point of left ventricular failure in end-stage, severe aortic stenosis, at which point the failing ventricle dilates. Myocardial hypertrophy of the left ventricular wall is recognizable by (*a*) widening of the interventricular septum as seen in the LAO view on composite views selected to demonstrate both ventricular cavities simultaneously; (*b*) displacement to the right of the septal border of the right ventricular cavity, due to septal hypertrophy; (*c*) increase in the space between the left ventricular cavity and activity in the left lung field; and (*d*) marked reduction in the apparent size of the left ventricular chamber secondary to hypertrophy and increased ejection fraction. (*2*) Poststenotic dilatation of the proximal ascending aorta is commonly noted.

These findings are illustrated in Figures 52 and 53, which show two studies on the same patient performed 16 months apart. The calculated aortic valve area from catheter studies was about 0.7 cm^2. Unfortunately, the patient died suddenly 2 days

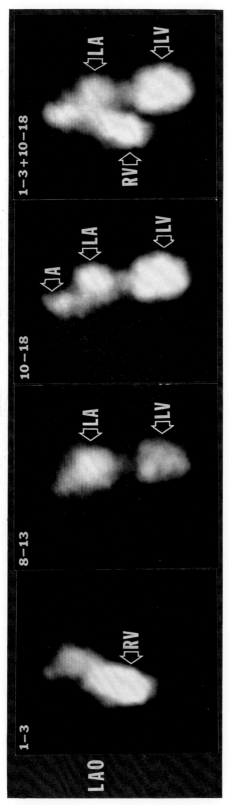

Fig. 51. Left ventricular aneurysm, oblique position. Note markedly dilated LV (frame 3) and prominent but not very large LA (frames (2–4), probably due to associated slight mitral insufficiency.

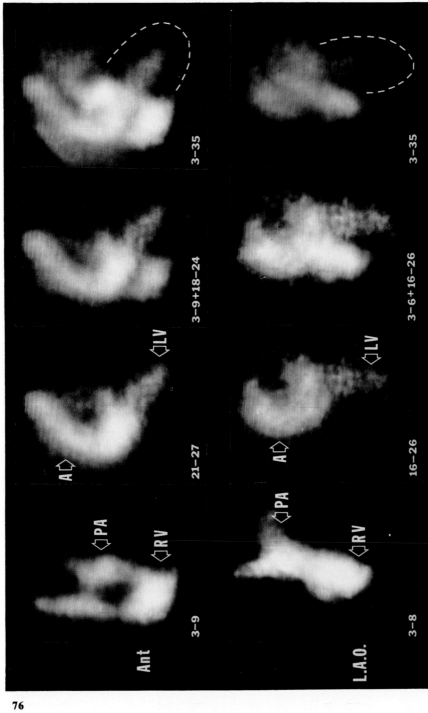

Fig. 52. Aortic stenosis, anterior and oblique positions. Note prolonged circulation time, relatively small LV cavity (frames 3 and 4, top), wide void separating radioactivity in LV from that in upper part of left lung (frame 4, top), widened ascending aorta. The dotted lines outline the estimated outer limits of the LV wall.

Valvular Heart Disease

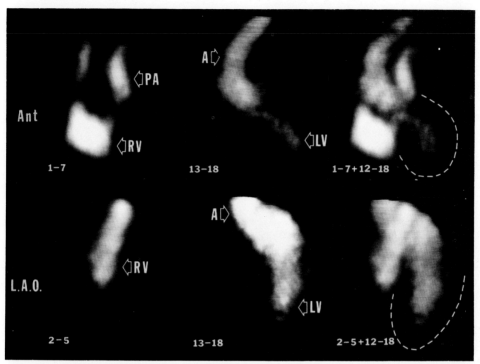

Fig. 53. Aortic stenosis, anterior and oblique positions. Same case as Figure 52, 16 months later. Note shift of interventricular septum to the right (frame 1, anterior), small-size cavity of LV (frames 2 and 3, anterior), and poststenotic widening of aorta. The dotted lines indicate the estimated outer limits of the LV wall.

prior to scheduled aortic valve surgery. There had been only minimal clinical indications that his condition had deteriorated in the interval between the two studies.

In aortic stenosis, the cavity of the LV may become so small that it is even difficult to visualize (Fig. 54). A similar finding is encountered in idiopathic subaortic stenosis (Fig. 55) as discussed in the following section.

The effects of surgical correction of aortic stenosis are shown in Figure 56. The changes observed were less dramatic than those following corrective surgery for mitral disease (Figs. 46 and 47) or valve replacement for aortic insufficiency (Figs. 57 and 58).

Subaortic Stenosis

In the few cases of documented subaortic stenosis (idiopathic hypertrophic form) which have been studied, findings were observed that were similar but not identical to those with valvular stenosis: (1) extremely small size of the left ventricular cavity which has a narrow, wedge shape with its apex pointed toward the ventricular apex; (2) little or no proximal aortic dilatation; and (3) moderate dilatation of the left atrium secondary to concomitant mitral insufficiency which is characteristic of this form of myocardopathy. These findings are illustrated in Figure 55.

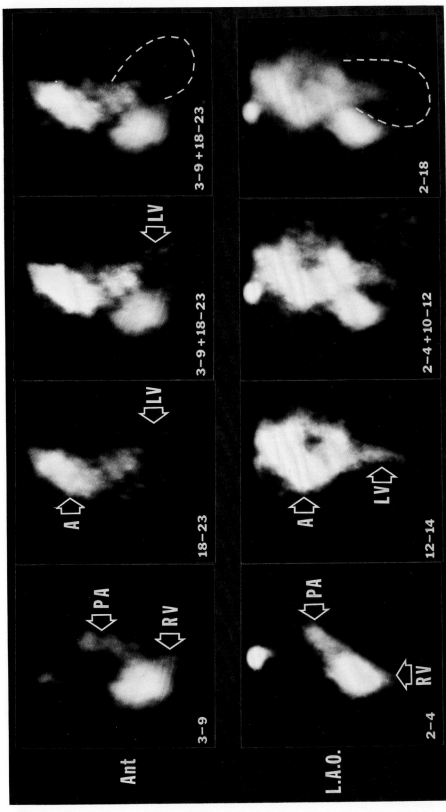

Fig. 54. Aortic stenosis, anterior and LAO positions. Note deviation of interventricular septum to right (anterior frame 1), small LV cavity, and poststenotic dilation of aorta (anterior and LAO frames 2–4). The dotted lines indicate the estimated outer limits of the LV myocardial wall.

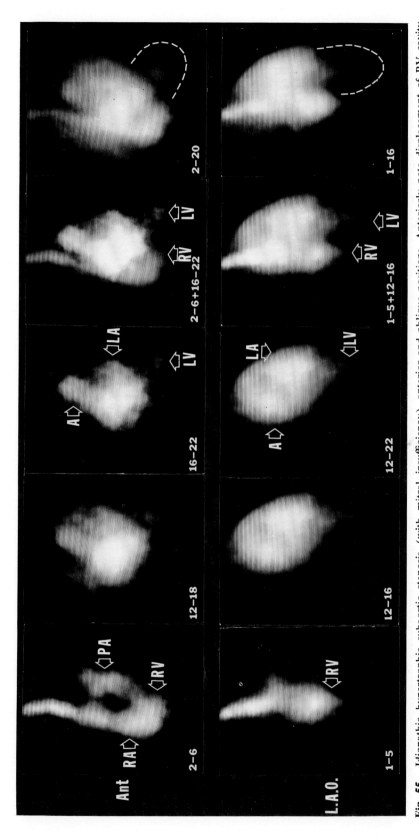

Fig. 55. Idiopathic hypertrophic subaortic stenosis (with mitral insufficiency), anterior and oblique positions. Anteriorly note displacement of RV cavity to the right (frame 1) due to thickening of interventricular septum, large LA (frame 3), small LV (frames 4 and 5), and wide space between RV and LV cavities representing hypertrophied interventricular septum. In oblique view note large LA (frames 2–5), small, wedge-shaped LV (frames 2–5), and wide separation between RV and LV cavities (frames 4 and 5) representing thickened interventricular septum. Aorta is well visualized in both positions. Dotted lines on last frames indicate estimated outer limit of LV wall.

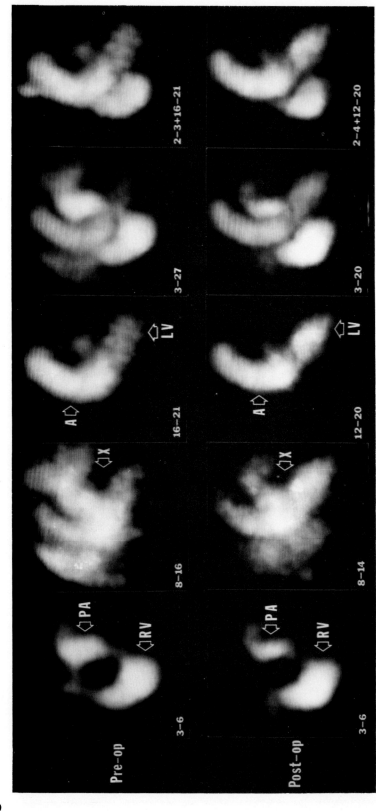

Fig. 56. Aortic stenosis, anterior position, pre- and postoperatively (2 months). In both studies note evidence of LV hypertrophy by shift of interventricular septum to the right (frames 1), wide separation between LV and cavity and left-lung activity (X, frame 2), and wide separation between RV and LV cavities (preop, frames 4 and 5). Poststenotic dilatation of aorta is noted preoperatively in frame 3. Postoperatively there was no significant change in bolus transit, but the LV cavity was more readily delineated; evidence of LV hypertrophy persists.

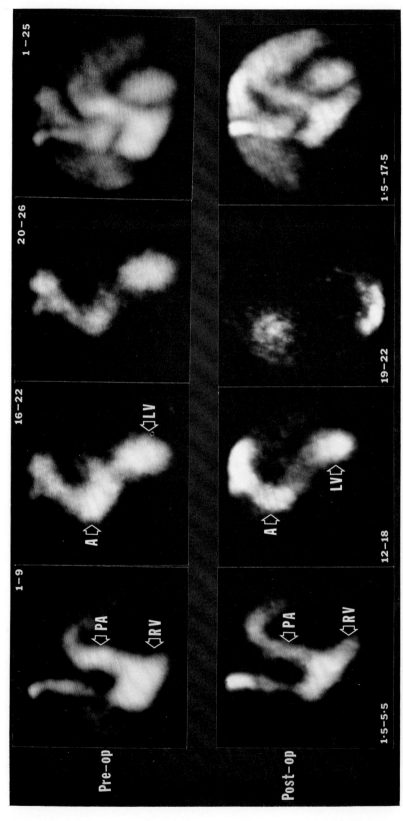

Fig. 57. Aortic insufficiency, anterior pre- and postoperative. In the preoperative series note the large left ventricle, prominent delineation of aortic valve region, and prolonged visualization of the left ventricle and aorta (noted also beyond 26 seconds in frames not shown). One week after aortic valve replacement, note the decrease in the heart size (both ventricles), faster circulation time, good emptying of the left ventricle (compare frame 3), and less clear delineation of the aortic valvular area.

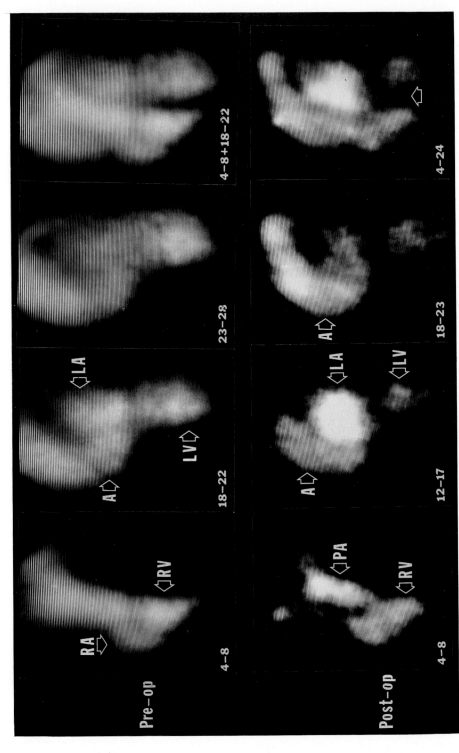

Fig. 58. Aortic insufficiency and mitral insufficiency, oblique position, before and 1 week after operation. Preoperatively note very large heart (frame 4), prolonged circulation time, very large left ventricle, large left atrium, and wide aorta (frames 2 and 3). Postoperatively (mitral and aortic valve replacement) note striking decrease in heart size, especially the left ventricle, persistent enlargement of left atrium, which empties better (frames 2 and 3), and improved circulation time. The width of the septal region (frame 3, arrow) suggests that left ventricular hypertrophy is present, made manifest by the elimina-

Valvular Heart Disease

Aortic Insufficiency

Analogous to the situation in which there is regurgitant flow through a mitral valve, incompetence of the aortic valve imposes an increased volume load on the left ventricle and, to a lesser extent, the aorta. The compensatory response to volume work is dilatation. The features of aortic insufficiency are: (*1*) left ventricular dilatation, often marked; (*2*) prominence of the ascending aorta reflecting its dilatation; (*3*) prolonged visualization of both left ventricle and aorta during the late phase of the study, with about equal intensity in the two regions; (*4*) clear identification of the location of the aortic valve as a region of decreased activity; and (*5*) vigorous ventricular contractions frequently observed on the television screen as the left ventricle repeatedly empties and refills secondary to regurgitant flow across the incompetent aortic valve.

Following surgical replacement of a diseased aortic valve, there is commonly noted a rapid, obvious reduction in the size of the left ventricular chamber, reduced circulation time, and disappearance of feature *3* noted above.

Scintigraphic findings typical of aortic insufficiency are shown in Figure 57. See also Figure 59, in which aortic insufficiency is associated with an aortic aneurysm. The effects of valve replacement in patients with this disorder may be dramatic (Figs. 57 and 58).

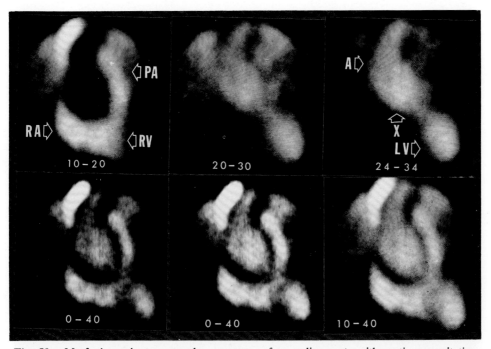

Fig. 59. Marfan's syndrome: saccular aneurysm of ascending aorta with aortic regurgitation, anterior position. Note widened aortic "window frame," (frame 1), dilatation of proximal aorta (*X*, frame 3), and LV enlargement (frame 3). In later frames not shown, LV and aorta remained visualized. Frames 4–6 show well the anatomic relationships of right- and left-sided cardiovascular structures.

Tricuspid Insufficiency

In tricuspid insufficiency, usually itself secondary to right ventricular failure and dilatation, there is extremely slow bolus transit, marked right atrial enlargement, and dilatation of the superior vena cava. In the case shown in Figure 60, right ventricular dilatation or hypertrophy could not be demonstrated due to poor bolus transit through the huge SVC–RA dilution chamber.

In patients with tricuspid insufficiency due to rheumatic heart disease, some degree of tricuspid stenosis is often also present. The authors have not had the opportunity to study a patient with predominantly tricuspid stenosis.

Pulmonic Stenosis

(See in Chapter 3 under Congenital Heart Disease)

Pulmonic Insufficiency

Only one example of this condition has been studied by the authors. The findings of prolonged residence of the bolus in RV and PA and right ventricular dilatation (Fig. 61) were those that might be expected from knowledge of the hemodynamic abnormality associated with to-and-fro movement across an incompetent pulmonic valve.

Mixed Valvular Lesions

The hemodynamic consequences of mixed mitral stenosis and insufficiency are operative over a continuous spectrum depending upon the relative degree of stenosis and insufficiency. Consequently, the angiocardiographic findings seen in combined stenosis and insufficiency range from that described for mitral stenosis on the one hand to that described for pure insufficiency on the other.

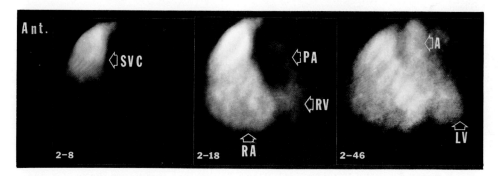

Fig. 60. Tricuspid insufficiency, anterior view. Note markedly dilated SVC (frames 1 and 2), huge RA (frames 2 and 3), and very poor bolus transit through the right heart.

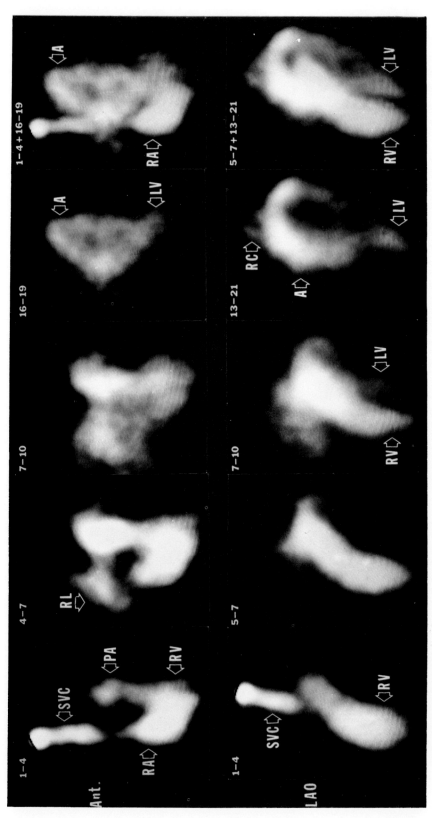

Fig. 61. Pulmonic insufficiency, anterior (top) and oblique (bottom) positions. Note prolonged residence in RV and PA in both studies, and RV dilatation, especially noticeable in the oblique position (frames 2 and 5). The study helped exclude other diagnoses which were being considered, such as aortic insufficiency and left-right shunt.

As in combined stenosis and insufficiency of the mitral valve, coexistence of obstruction and regurgitation in the aortic valve can present with angiocardiographic manifestations of both lesions. Accurate assessment of the relative magnitude of stenosis or insufficiency is not possible using qualitative methods alone. Methods of data processing with the aid of a digital computer designed to permit such assessment are discussed in Chapters 7 and 8.

Various combinations of mixed stenosis and insufficiency of the aortic and mitral valves are commonly seen and the angiocardiographic features of both lesions may be combined. Examples of a number of different combinations are shown, as follows:

Figure 62, combined mitral stenosis and aortic stenosis.
Figure 63, combined mitral insufficiency and tricuspid insufficiency.
Figure 58, combined aortic insufficiency and mitral insufficiency.
Figure 64, combined mitral stenosis, aortic stenosis, and aortic insufficiency.
Figure 49, combined mitral stenosis and mitral insufficiency.

The angiographic data in the absence of other clinical information can be difficult to interpret, but certain generalizations can be made: (*1*) dilatation of the left atrium indicates at least mitral valvular dysfunction; (*2*) dilatation of the left ventricle and ascending aorta is very suggestive of at least aortic valvular disease; (*3*) prominence of the left atrium with reduction in size of the left ventricular cavity can be seen in combined mitral valvular disease and aortic stenosis (see previous discussion of subaortic stenosis); (*4*) in the presence of combined aortic insufficiency and mitral valvular disease, the angiographic study may incorrectly present features suggesting pure mitral insufficiency, if there is absence of marked prominence of the ascending aorta.

Much of the diagnostic difficulty can be minimized if one takes into account other important information known about the patient, particularly the findings on physical examination of the heart together with electrocardiographic and conventional radiographic cardiac series.

It should be emphasized in this connection that as yet we do not have absolute criteria for diagnosing "enlargement" or "dilatation" of a given chamber, nor of "thickening" of a ventricular wall. Evaluation of these parameters is qualitative, and is made after noting the size of the heart image in its entirety and after comparing relative chamber and wall thickness sizes in the positions studied in any given case. Gross disparities from the normal are not difficult to recognize; on the other hand, slight but clinically significant abnormalities could be missed. It should be emphasized that valuable additional information relative to transit time, residence time, dynamics of ventricular contraction, anatomic contours, and position relationships of the cardiac chambers are also obtained during a study, and this information should be taken into account before rendering a diagnostic opinion. The interpretation of qualitative data, supplemented by clinical information, involves a diagnostic process not dissimilar to that employed by the diagnostic roentgenologist. Diagnostic accuracy will undoubtedly be greatly improved when quantitative criteria of normality can be applied to the various parameters derived in this type of examination.

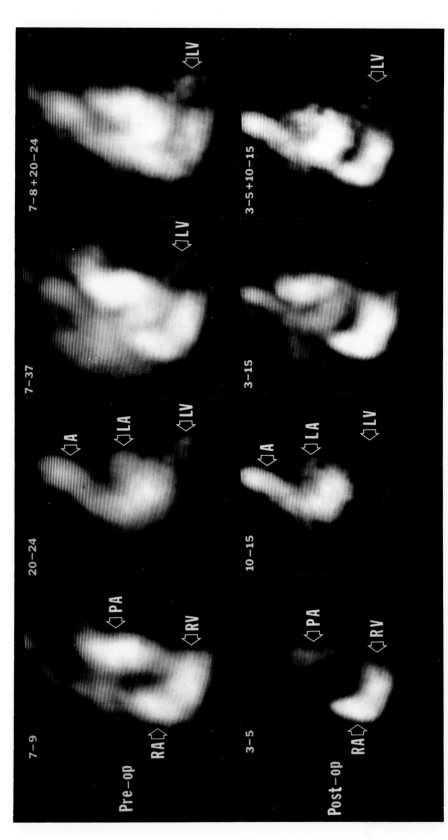

Fig. 62. Combined mitral stenosis and aortic stenosis, anterior view, pre- and postoperatively (2 months). Preoperatively note large LA (frame 2), small LV cavity (frames 2–4), and poststenotic dilatation (frame 2). Postoperatively note decreased heart size (all frames), smaller LA, persistently small LV cavity, and faster bolus transit.

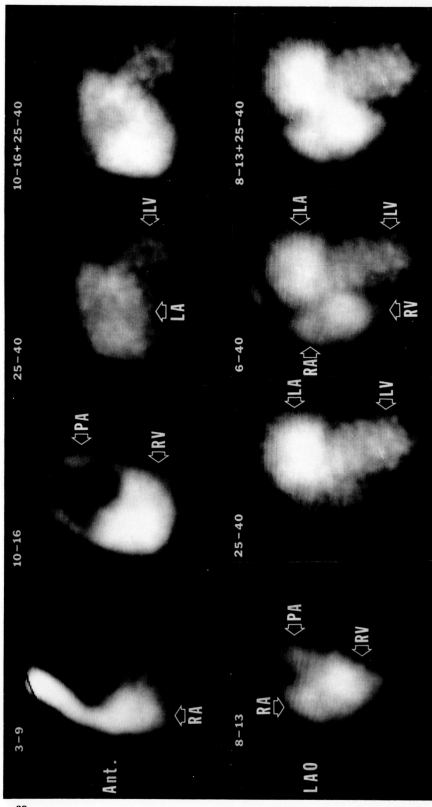

Fig. 63. Combined mitral insufficiency and tricuspid insufficiency, anterior (top) and oblique (bottom) positions. Note large RA (anterior, frames 1 and 2), marked prolongation of bolus transit, probable enlargement of RV (oblique, frame 1), very large LA (anterior, frame 3; oblique, frames 2–4), and dilated LV (anterior, frames 3 and 4; oblique, frames 2–4).

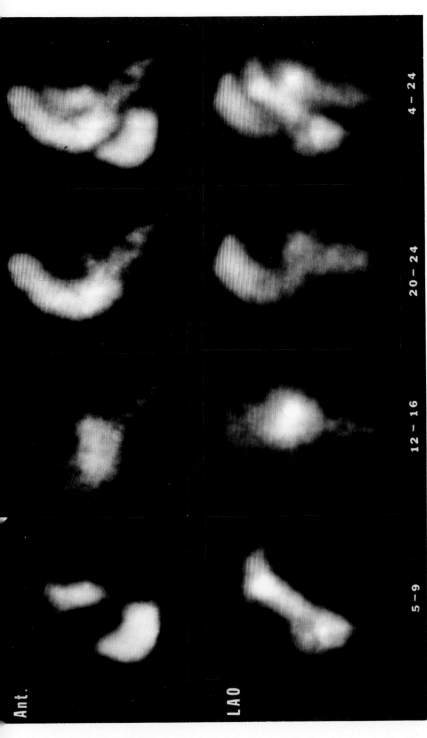

Fig. 64. Combined mitral stenosis, aortic stenosis, and aortic insufficiency, anterior (top) and oblique (bottom) positions. Anteriorly note signs of LV hypertrophy: deviation of interventricular septum to the right (frame 1) and abnormal separation of RV and LV cavities (frame 4); note also LA enlargement (frame 2) and dilatation of aorta (frame 3). In oblique position, note enlarged LA (frames 2 and 4) and wide IV septum (frame 4). Despite the mitral stenosis, the LV and aorta were well visualized for a prolonged period. This finding is unusual in cases of pure mitral stenosis and is consistent with diagnosis of aortic regurgitation. In addition, the time–activity curve over the LV showed abnormally large activity differences during diastole and stystole. A gelatinous clot was also found in the LA at operation.

Left Ventricular Hypertrophy

Extreme left ventricular hypertrophy without valvular lesions and without evidence of proximal aortic dilatation or left atrial dilatation may occur in patients with long-standing severe hypertension, for example, due to chronic renal disease or essential hypertension. In the absence of congestive failure, the radionuclide angiographic features in such cases are limited to the signs of left ventricular hypertrophy cited previously (see discussion under Aortic Stenosis). A typical example of concentric hypertrophy of the LV associated with severe essential hypertension is shown in Figures 65a and 65b.

Aneurysms

Left Ventricular Aneurysm

While the diagnosis of some ventricular aneurysms may be suggested by electrocardiography, radiography, or cinefluoroscopy, the firm diagnosis of an aneurysm, together with its anatomic location and possible paradoxical myocardial kinetics, is usually established by conventional left ventricular contrast angiography. However, radionuclide angiocardiography has been useful as a screening test in detecting this lesion and is probably underused for this purpose.

The normal left ventricular cavity appears in the usual radionuclide study as an elongated or oval shape, sometimes tapering at its distal extent, and with a width usually not exceeding that of the normal aortic root. (Due to the fact that framing time is usually several seconds long, the left ventricular cavity image is delineated by its limits during diastole. Special framing at short intervals is required to obtain separate pure systolic and diastolic images.) Aneurysms may appear on the study as focal bulges in various portions of the ventricle; there is usually prolonged residence of radioactivity in the involved region (14). The ventricle as a whole may be dilated, and the accuracy of diagnosis usually depends on the ability to demonstrate, on at least one view, a localized region where the aneurysmal bulge is especially noticeable. However, unless the aneurysm is large or is border-forming on one or both of our two conventional views, it may remain masked during routine studies. Rotation of the patient to an optimal angle to record a view tangential to the aneurysm increases the sensitivity of the method.

To illustrate the variety with which LV aneurysm may present, several examples are presented in Figures 66–70. The effects of surgical repair (10) of the aneurysm are well shown in Figures 66, 67, and 70, while Figure 68 demonstrates the use of radionuclide angiocardiography to monitor the development of aneurysm by performing serial studies in an affected patient.

Zaret et al. have studied regional left ventricular dysfunction with scintiphotographic techniques following the intravenous administration of labeled albumin and subsequent precordial imaging over a relatively long period using special timed exposures (14). By gating the scintillation camera exposures from an electrocardiographic signal, one can obtain separately summed end-systolic and end-diastolic images that can be compared for changes in ventricular cavity dimensions. Areas of

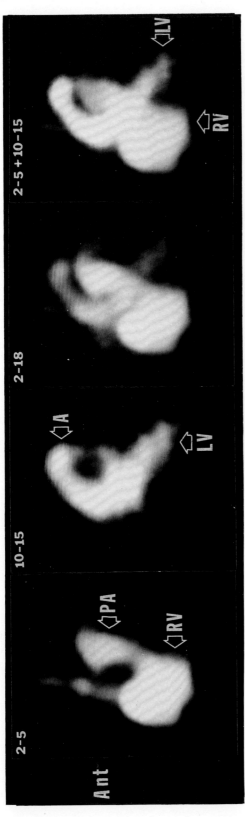

Fig. 65a. Left ventricular hypertrophy due to severe essential hypertension, anterior position. Note shift of interventricular septum to right (frame 1) and wide separation of RV and LV cavities (frames 3 and 4). Bolus transit is normal. The study was performed to rule out a possible atrial septal defect, which is easily excluded by the absence of early recirculation of tracer through the right atrium.

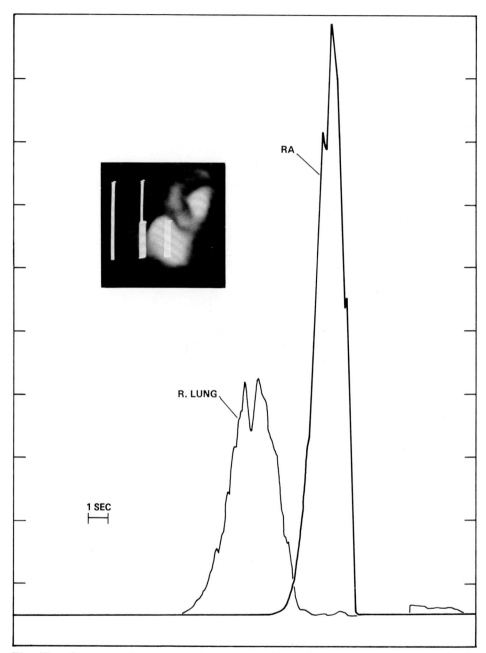

Fig. 65b. Curves over RA and right-lung regions of interest showed normal patterns, that is, time–activity curves over RA and right lung showed typical normal sharply rising and falling activities in each region.

Valvular Heart Disease

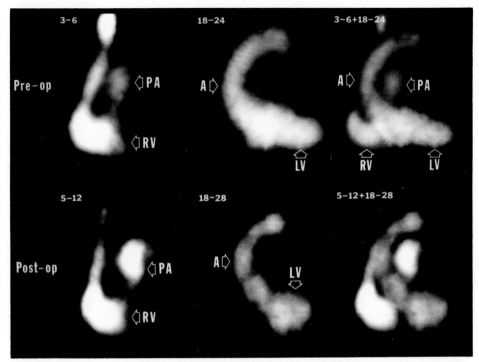

Fig. 66. Ventricular aneurysm, anterior position, before and after (1 month) operation. Preoperatively, note the irregular gross dilatations of the left ventricle with aneurysmal ballooning of the distal portion (frames 2 and 3) and the prolonged circulation time. Postoperatively, the left ventricle is much smaller because of excision of the aneurysm, but the speed of the circulation is unchanged.

akinesis are shown by failure of a given region of left ventricular wall to move during cardiac contraction.

We have employed a digital computer to assess regional left ventricular contraction in dynamic mode during the first pass of radiopharmaceutical through the heart. By framing the data at 10 to 20 frames/second, one can obtain time–activity curves over different selected regions of the ventricle. Paradoxical motion in a given region, for example the apex, is detected by showing that the time–activity curve of the region is out of phase with that of another normal region (Figure 71). Thus, while the scintiphoto may suggest the presence of aneurysm, computer analysis can graphically demonstrate relative hypokinesis, akinesis, or paradoxical motion.

Aneurysm of the Sinus of Valsalva

The successful demonstration of an aneurysm of the sinus of Valsalva (10), with postoperative assessment of its repair is shown in Figure 72.

Aneurysm of the Aorta

Aneurysms of the ascending aorta present as localized bulges of the proximal aorta just beyond the aortic valve. They may be well visualized in the anterior position, and there is not infrequently an associated aortic insufficiency (Figs. 73 and 59).

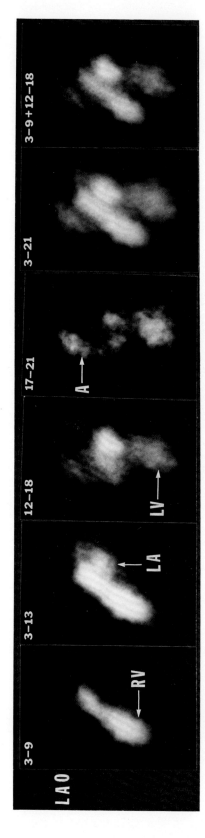

Fig. 67. Postoperative repair of LV aneurysm (1 year), LAO position. The left atrium (LA) is well visualized (frames 2, 3, 5, and 6) even though its size is normal. This finding has been noted in several other cases with LV aneurysm and may be due to associated papillary muscle dysfunction and slight mitral insufficiency. The LV is still slightly dilated, and the circulation time slightly prolonged

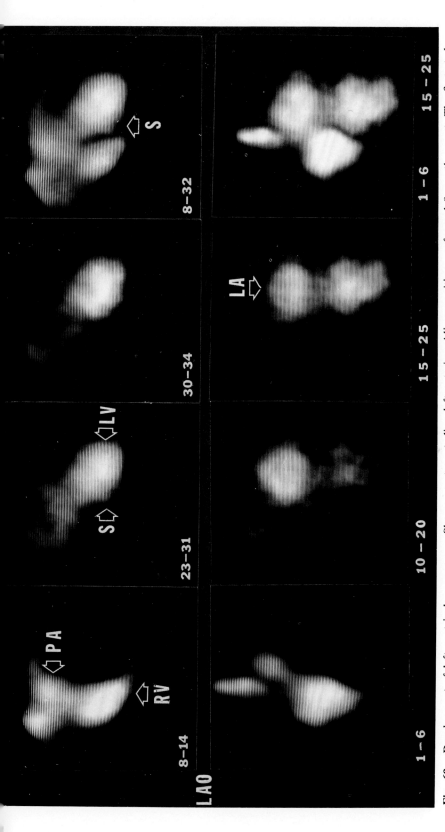

Fig. 68. Development of left ventricular aneurysm. Shown are two studies, left anterior oblique position, performed 8 months apart. The first study (top) was performed a few days after an acute antero-septal myocardial infarct in order to evaluate a possible rupture of the septum (acquired VSD). Note the bulge in the septum (5, in frames 2 and 4), but no evidence of VSD (absence of RV activity during filling of LV, frame 3). The bolus transit is very prolonged. The second study (bottom) shows irregular bulging of the LV, not only in the septal region, but distally (frames 3 and 4), and enlargement of the LA has developed (frames 2–4).

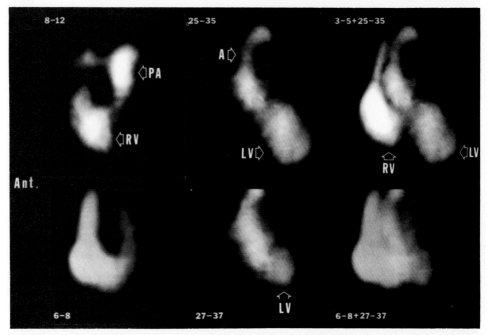

Fig. 69a. Left ventricular aneurysm, anterior position, preoperatively (top) and 1 week postoperatively (bottom). Preoperatively note deviation of interventricular septum to the right (frame 1) and very large LV cavity (frames 2 and 3). The width of the interventricular septum is normal (frame 3). Postoperatively (resection of aneurysm) note different contour of right heart (frame 1) and smaller truncated LV.

Aneurysms of the descending aorta are less easily seen in an anterior study because of the distance of the descending aorta from the detector and the diminished radioactivity in it. To detect an aneurysm of the descending aorta, the patient should be studied in the left anterior oblique position or in a posterior oblique position. A dramatic example of a saccular aneurysm of the descending aorta is shown in Figure 74.

Dissecting Aneurysm of the Aorta

The author has had relatively little experience studying patients with this condition. The diagnosis may depend on demonstration of a double-lumened widened aorta, and the resolution of the scintillation camera is probably inadequate for such a purpose when injections are made on the right side of the central circulation (Fig. 75).

Luetic Aortitis and Aneurysm

Figure 76 demonstrates irregular bulging in the proximal aorta, but also gross irregularities of contour in the arch and descending aorta in a patient with postmortem evidence of severe luetic involvement of the aorta.

Valvular Heart Disease

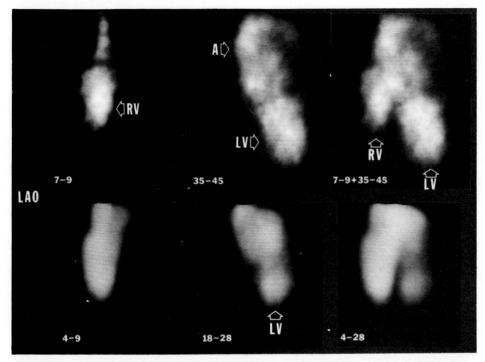

Fig. 69b. Left ventricular aneurysm, modified left anterior oblique position, preoperatively (top) and 1 week postoperatively (bottom), same case as in Figure 69a. Preoperatively note very large LV (frames 2 and 3) and prolonged bolus transit. Postoperatively (resection of aneurysm) note much smaller LV.

Aneurysm of the Pulmonary Artery

Aneurysm of the pulmonary artery should be easily diagnosed by radionuclide angiocardiography because the lesion is relatively close to the bolus injection site. In the one case the author has studied, striking abnormalities of the pulmonary artery were visualized in both anterior and oblique positions (Fig. 77).

Redundant (Uncoiled) Aorta

Redundancy of the aorta may cause contour changes in the chest roentgenogram which resemble those observed in patients with aortic aneurysm. In the striking example shown in Figure 78, the radionuclide study was considered adequate to exclude the latter diagnosis.

Cardiomyopathy

The angiocardiographic findings in cardiomyopathy are suggestive, but are not conclusive, for this diagnosis. The following abnormalities have been noted: (*1*) four-

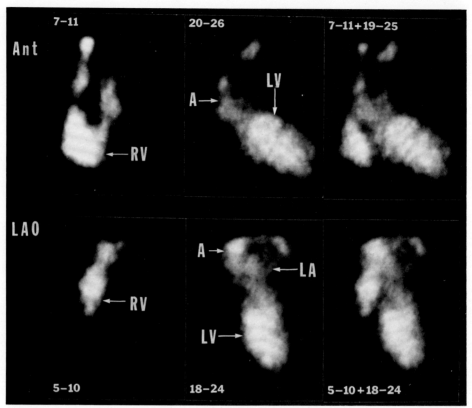

Fig. 70. Left ventricular aneurysm, anterior (top) and oblique (bottom) positions. Note gross and somewhat irregular dilatation of the LV (both studies, frames 2 and 3).

chamber cardiac dilatation; (2) marked prolongation of transit times of all phases of the study; (3) absence of other signs of cardiac pathology, such as left-to-right shunt, isolated valvular disease, and absence of ventricular hypertrophy. Figure 79 gives the findings observed in an adult; Figure 80 shows the findings in a child.

Similar findings may be noted in individuals with end-stage mitral insufficiency or in individuals with late-stage congestive failure of any cause, in which the physiologic defect of large volumes and low flow are found. Often the clinical history or previous observations of the patient enable the physician to exclude the latter possibilities.

Coronary Artery Disease

If coronary artery disease is not associated with LV aneurysm, LV dilatation, or congestive heart failure, the patient with angina pectoris due to coronary artery disease may have no discernible abnormality when tested by intravenous radionuclide angiocardiography, either before or after coronary artery bypass grafting (Fig. 81).

When an anatomically normal heart is transplanted in place of a severely damaged one, then pre- and postoperative studies on the same subject reveal grossly

Valvular Heart Disease

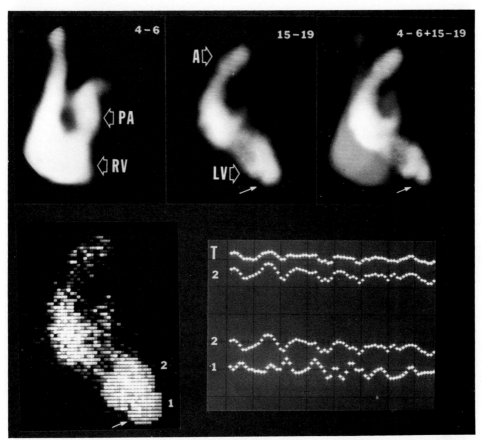

Fig. 71. Left ventricular aneurysm with paradoxical pulsation, anterior view, with computer-generated time–activity curves. Note the small LV apical bulge (frames 2 and 4, top). The arrows point to a bulge in the distal left ventricle. Frame 1 (bottom row) demonstrates the computer-generated image corresponding roughly to frame 2 (above). It has two selected regions of interest: area *1* in the region of the apical bulge and area *2* the remainder of the LV. Frame 2 (bottom) represents time–activity curves of regions *1* and *2* (and total, *T*) during bolus transit through the LV, processed at 20 frames/second, and statistically smoothed. Comparing curves 2 and 1, one can see that they are out of phase most of the time throughout a series of six cardiac cycles.

different anatomic and hemodynamic features (Fig. 82). The function of the grafted heart can be assessed with very little difficulty by performing repeated radionuclide studies (see Chapter 5, left ventricular performance).

Cardiac Tumors

Left Atrial Myxoma

Striking findings were seen preoperatively in two patients with left atrial myxoma who presented with clinical and auscultatory signs similar to those seen in mitral

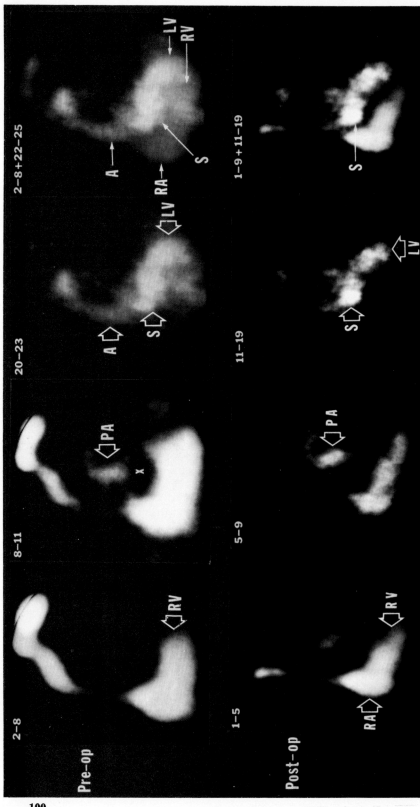

Fig. 72. Aneurysm of sinus of Valsalva, producing right ventricular obstruction, anterior position. Preoperatively note very large right atrium and ventricle (frames 1 and 2), region of obstruction (X) to the right ventricular outflow (frame 2), delayed right-heart emptying, and dilatation near aortic root with increased density at its lower portion (S) in frame 3. Composite view (frame 4) shows that S lies in region corresponding to right ventricular outflow obstruction (X, frame 2). Four months after operation note greatly decreased size of right heart (frame 1), diminished size of sinus of Valsalva aneurysm (S), with relief of the previously obstructed right ventricular outflow and shorter circulation time.

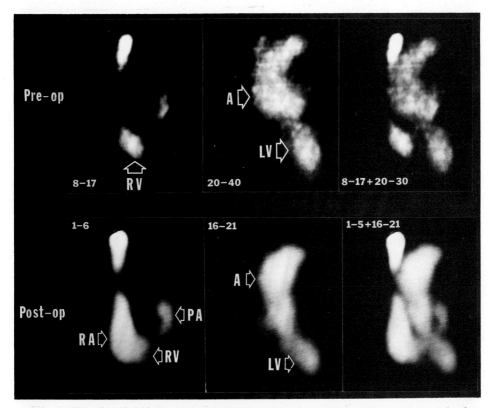

Fig. 73. Aortic aneurysm and aortic insufficiency, anterior position, before and 1 week after operation. Preoperatively note saccular enlargement at root of aorta (frames 2 and 3), prominent LV cavity, clearly delineated aortic valvular area, widening of aortic area between superior vena cava and pulmonary artery (frame 1), and prolonged circulation time with late dual visualization of LV and aorta. Postoperatively (aortic valve replacement and aneurysm repair) note smaller LV, absence of aneurysm at aortic root, poorer delineation of aortic valvular area, shorter circulation time, and widening of ascending aorta near the arch.

stenosis (Figs. 83 and 84). In both instances echocardiography and contrast cineangiocardiography demonstrated mass lesions in the left atrium. In the case shown in Figure 83, the mass was a spherical, pedunculated tumor which produced intermittent obstruction through an otherwise normal mitral valve. In the case shown in Figure 84, there was prolapse of the tumor through the valve. The radionuclide angiocardiogram in patients with left atrial myxoma has features which resemble those of mitral stenosis, but in addition shows evidence of a filling defect within the LA. It is possible that a similar set of findings might occur in patients with mitral stenosis who develop large left atrial clots.

Metastatic Tumor

Figure 85 demonstrates a filling defect in the RV due to invasion of the heart by a metastatic carcinoma of the colon.

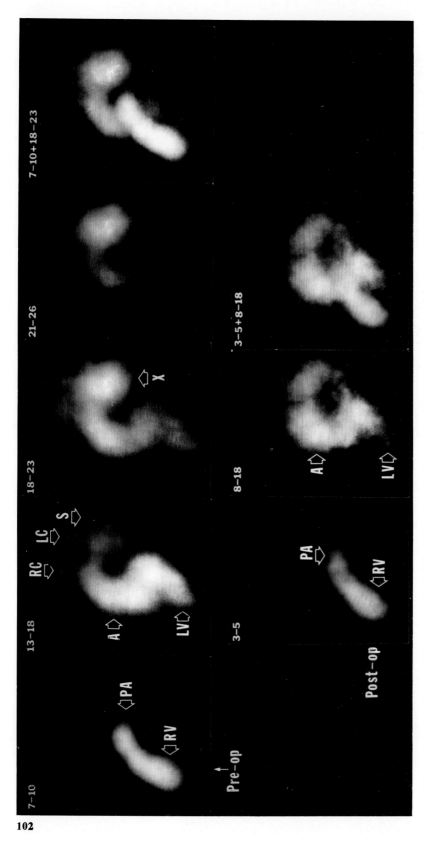

Fig. 74. Aneurysm of descending aorta, oblique position, before and 1 week after operation. Preoperatively note saccular aneurysm of descending aorta (X, frames 3–5) at or near takeoff of left subclavian artery (S, frame 2). RC, right common caroted artery, and LC, left common carotid artery. Postoperatively (excision and repair of traumatic aneurysm), the aneurysm can no longer be seen; there is an increase in width of the aorta proximal to the site of repair (frames 2 and 3).

Valvular Heart Disease

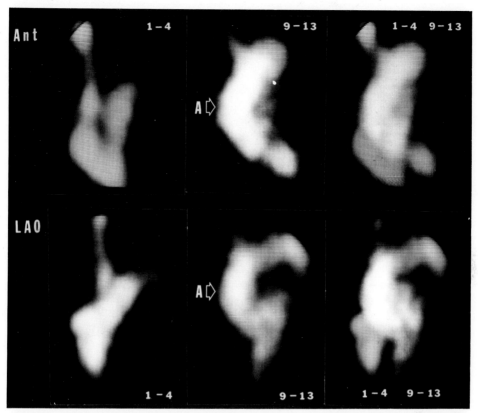

Fig. 75. Acute dissecting aneurysm of the aorta, anterior and oblique positions. Bolus transit was normal. The aorta was visualized well in both positions and was irregular and kinked in the oblique position. Although the "kink" in the aortic arch was one site of an extensive dissection, the study was interpreted as having failed to delineate the dissection (contrast angiography showed a double-lumened aorta.)

Essential Pulmonary Hypertension

Figure 86 illustrates the findings seen in patients with primary pulmonary hypertension (5):

1. Prolonged residence time of the bolus in the superior vena cava and right heart chambers.
2. Reflux of radioactivity into the innominate vein contralateral to site of injection.
3. Enlargement of right atrium and right ventricle.
4. Striking decrease in visualization of main right and left pulmonary arteries and lungs, presumably due to narrowing of these vessels.
5. Delayed visualization of left ventricle and aorta.

Similar findings might be found in patients with severe pulmonary hypertension due to some specific lesion, such as multiple pulmonary emboli or as a consequence of a long-standing left-to-right shunt, as shown in Figure 87. Thus, other tests would be necessary to establish the etiology of the pulmonary hypertension.

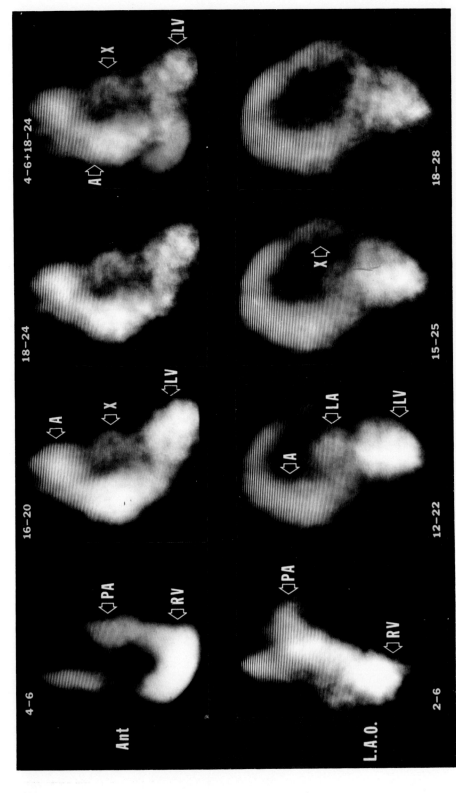

Fig. 76. Syphilis of the aorta, anterior and oblique positions. Note in both studies the dilated and grossly irregular and ragged contour of the ascending aorta (frames 2–4) and an aneurysmal bulge in the descending aorta (X, anterior frames 2 and 4 and oblique frames 3 and 4).

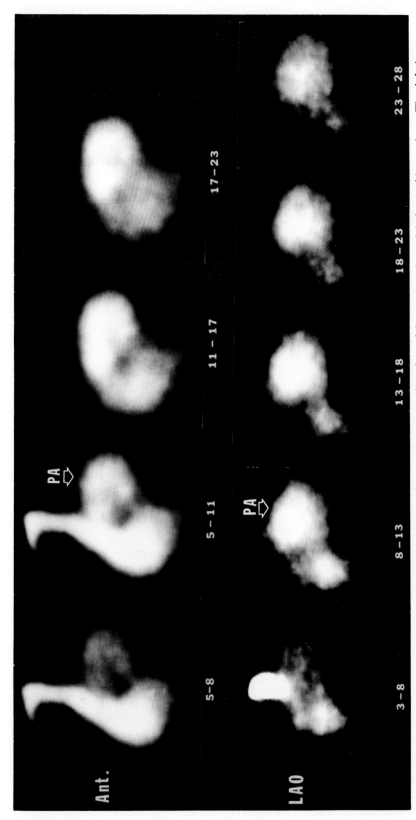

Fig. 77. Pulmonary artery aneurysm, anterior and oblique positions. Note gross dilatation of PA and prolonged bolus residence there. The left heart was never well seen.

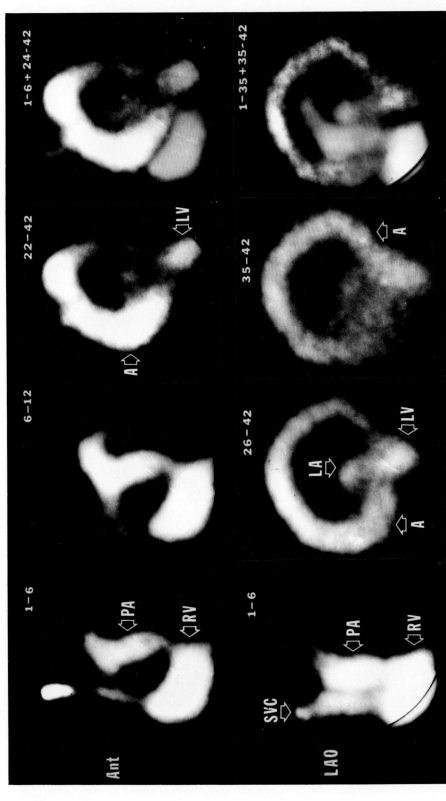

Fig. 78. Redundant aorta, anterior and posterior positions. Note marked widening of the aorta (anterior, frames 3 and 4; oblique, frame 2) and unusual course of this vessel (anterior, frame 4; oblique, frames 2–4). The study was performed to rule out a saccular aneurysm of the aorta suspected because of chest roentgenographic findings.

Valvular Heart Disease

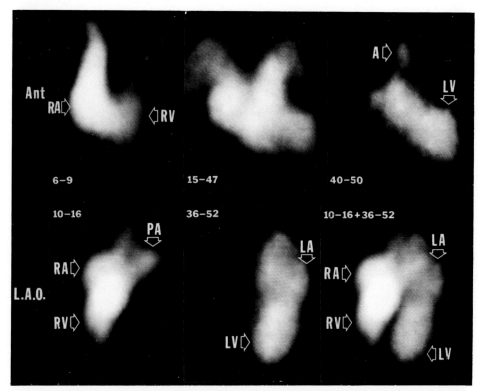

Fig. 79. Idiopathic myocardopathy, anterior and oblique positions. Note profound prolongation of circulation time, massive cardiomegaly involving all four heart chambers (e.g., frame 3, LAO), and absence of signs of left-to-right shunt (frame 3, anterior, and frame 2, LAO).

Hyperkinetic Cardiovascular States

In hyperthyroidism, either spontaneous or iatrogenic, one may observe evidence of rapid bolus transit in the absence of any other angiocardiographic abnormality (Fig. 88). Similar findings may occur in exercise, or in any instance of a hyperdynamic circulation such as adrenergic state or peripheral A-V fistula (Fig. 89). Fast transit from RV to LV might also be expected to occur acutely in circumstances where pulmonary blood volume is markedly reduced, for example, in massive pulmonary embolism (see Chapter 5, left ventricular function).

Superior Vena Cava Obstruction

Figure 90 shows the striking findings seen in this disorder (6), namely:

1. Demonstration of collateral veins carrying the radioactive bolus downward past the heart into the subdiaphragmatic region.
2. Absence of filling of the superior vena cava.
3. Late filling of the heart, apparently via inferior vena cava.
4. Prolongation of total circulation time, but normal right heart-to-aorta circulation time.

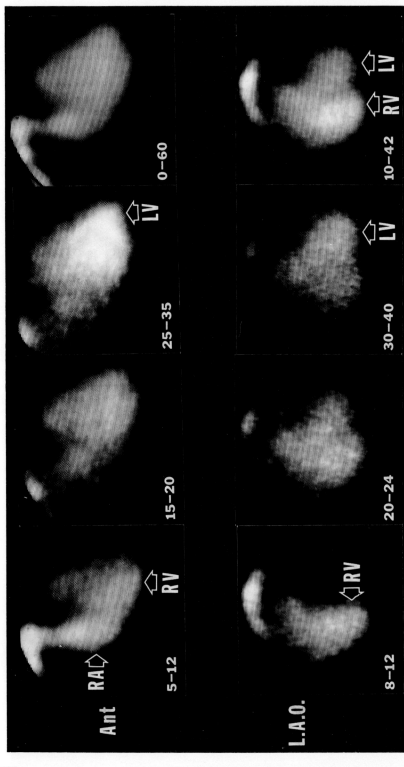

Fig. 80. Idiopathic cardiomyopathy in an 11-year-old child, anterior and posterior positions. Note very slow bolus transit and biventricular dilatation (anterior, frame 1; oblique, frame 4) and also RA enlargement; the LA was not well visualized. A distinct left-sided phase could not be obtained due to the slow bolus transit.

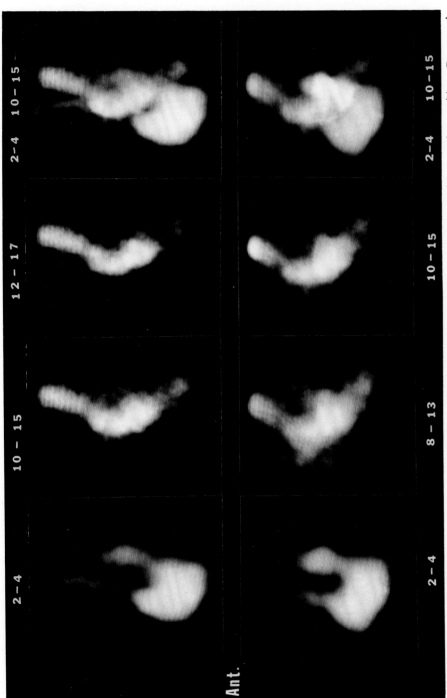

Fig. 81. Coronary artery disease, pre- and 1 week postoperatively (coronary artery by-pass graft), anterior position. Preoperative sequence is at top, postoperative sequence at bottom. Aside from minor configuration changes probably due to slight differences in positioning, there are no significant anatomic or hemodynamic differences between the two studies.

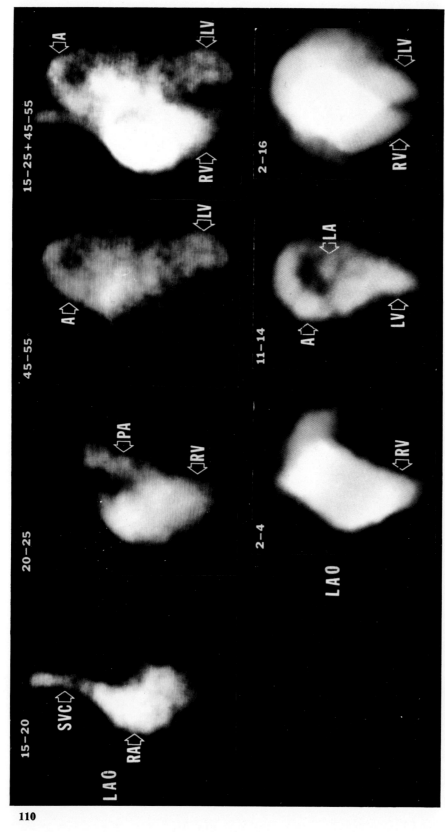

Fig. 82. Coronary artery disease and cardiomyopathy, before and after heart transplantation (4 months). Preoperatively (top) note profound slowing of bolus transit and gross enlargement of all heart chambers. Postoperatively (bottom) the transplanted heart at 4 months showed normal configuration and function.

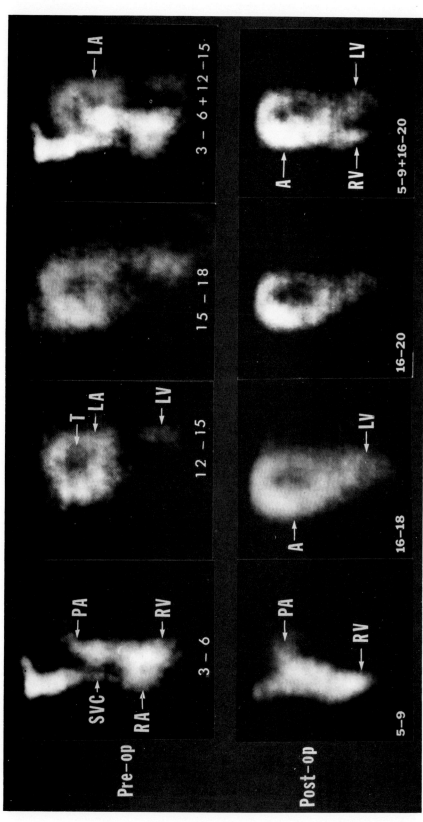

Fig. 83. Left atrial myxoma, oblique position, before and 1 week after operation. Preoperatively note enlarged LA (frames 2–4), filling defect (T) in LA (frames 2–4), and failure to delineate clearly the aortic outflow; frames between 18 and 30 seconds continued to resemble frame 3. A spherical myxomatous tumor about 5 cm in diameter was removed. Postoperatively, the angiographic pattern was normal; a possible reader's impression of filling defect in frames 2–4 is due to normal pattern of LV–aortic arch flow.

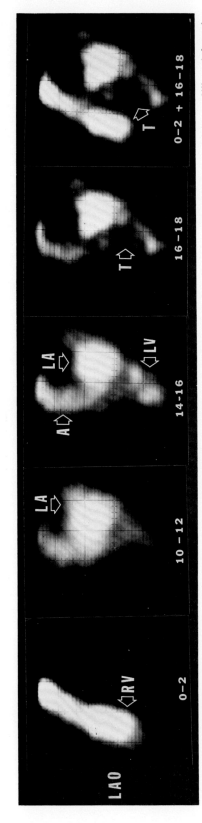

Fig. 84. Left atrial myxoma, oblique position. Note enlarged left atrium and prolonged bolus residence there (frames 2–4) and large filling defect due to tumor (T) involving medial portion of both LA and LV (frames 4 and 5).

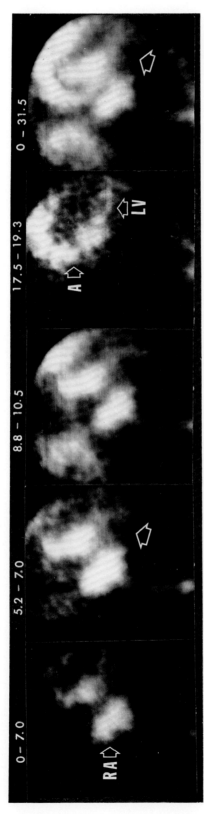

Fig. 85. Colon carcinoma invading right ventricle, anterior position. Injection was made via inferior cava catheter (catheter is seen best in frame 2). Filling defect of right ventricle is seen during right-sided filling (large arrow, frames 2 and 5). Defect is also seen on playthrough scintigraph (large arrow, frame 5).

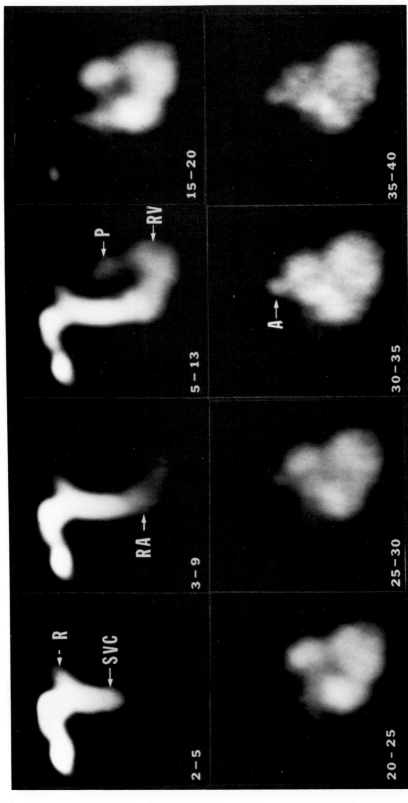

Fig. 86. Primary pulmonary hypertension, anterior view. Note reflux (*R*) into innominate vein contralateral to injection site, very slow movement through right heart, bulbous configuration of RV and prominent RA in frames 4–8, failure to delineate main pulmonary arteries, late visualization of aorta, and prolonged circulation time. P, pulmonary conus.

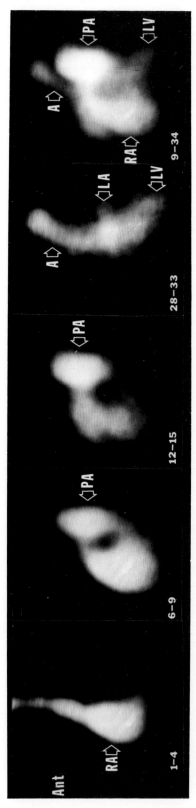

Fig. 87. Corrected atrial septal defect, 6 years following surgical closure of defect, anterior position. Note prominent RA (frame 1), relatively small-sized cavity at right ventricle (frame 2), dilated main pulmonary artery (PA, frames 2 and 3), prolonged circulation time through both right and left heart (frames 3 and 4), and striking prominence of PA on composite view (frame 5); findings are attributed to residual pulmonary hypertension.

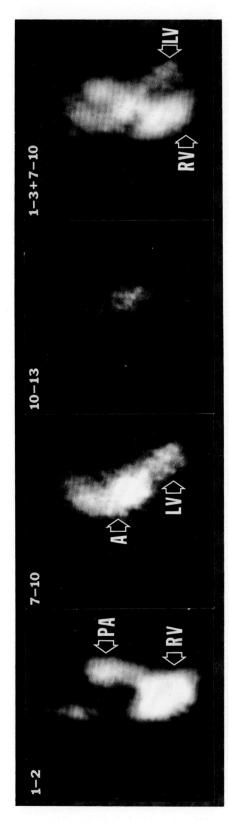

Fig. 88. Graves' disease, anterior position. Note very rapid bolus transit with LV and aorta visualized at 7 to 10 seconds, and rapid cardiac emptying at 10 to 13 seconds. Heart chambers and vessels are normal.

Valvular Heart Disease

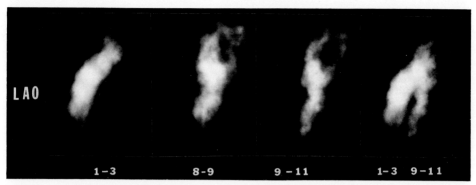

Fig. 89. Hyperkinetic heart in an adult, oblique position. Note very fast bolus transit, normal anatomic features. Although at the time of this study thyroid function tests were normal, the patient developed overt Graves' disease 3 years later.

An unusual case associated with pericardial effusion is shown in Figure 91. A single large vein apparently bypassed the obstructed vena cava in the case shown in Figure 92.

In cases with obstruction due to tumor treated by mediastinal radiotherapy, the test may be repeated at intervals to assess the effect of treatment in relieving the obstruction.

Pericardial Effusion

Dynamic radionuclide angiocardiography in patients with pericardial effusion shows the following diagnostic features (1, 2):

1. Wedge-shaped filling defect at the *right* cardiopulmonary border.
2. Wedge-shaped filling defect at the *left* cardiopulmonary border even after left ventricular filling.
3. Delineation of the pericardial sac as a void around the heart in later phases of the study.
4. Delay in bolus transit.
5. Frequent demonstration of an associated pleural effusion shown by failure to demonstrate complete encirclement of the pericardial sac by visualized radioactivity, especially on the left side.

Figure 93 demonstrates these diagnostic features and illustrates for comparison the findings observed when the enlarged cardiac silhouette is due to marked cardiac enlargement and congestive heart failure. Recent studies in the author's laboratory indicate that effusions larger than 70 ml can be reliably detected by this method. Echocardiography compares more favorably with respect to sensitivity, portability, and cost. The radionuclide study may be helpful when anatomic information is desired or when echocardiographic results are equivocal.

The diagnoses described in this chapter were in all instances established or confirmed by other diagnostic procedures, including phonocardiography, echocardiography, cardiac catheterization, roentgenography, and contrast angiocardiography,

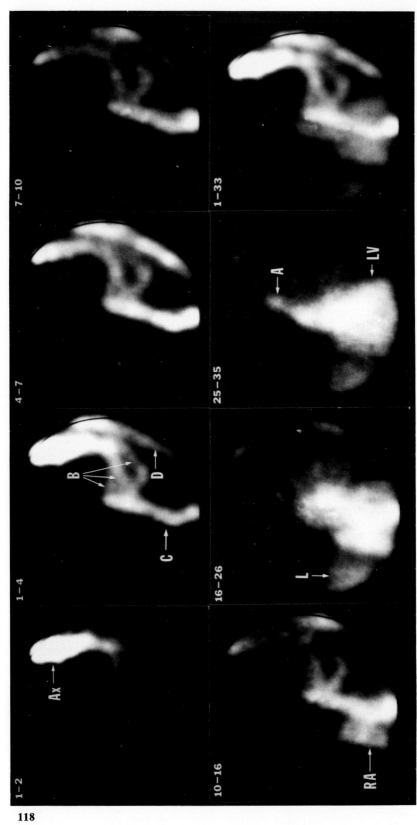

Fig. 90. Superior caval obstruction, anterior position. Injection was made into left antecubital vein. Note filling of left axillary vein (Ax, frame 1), intercostal collaterals (B, frame 2) axillary collateral vein (C, frame 2), and internal mammary vein (D, frame 2), but no superior vena caval or cardiac activity was seen up to 10 seconds (frames 1–4). RA filled from below (frame 5), and aorta and LV were visualized late (frame 7). The composite view (frame 8) shows well the relationship of the heart (faint shadow in the center) to the collateral venous system. The obstruction was caused by metastasis from a bronchogenic carcinoma.

Valvular Heart Disease

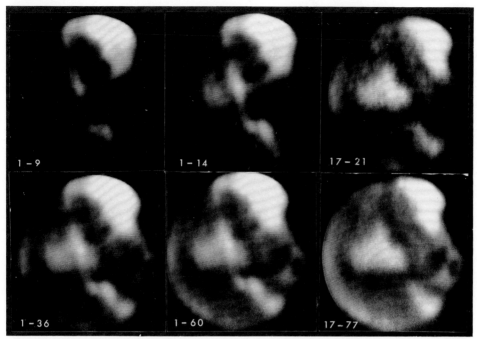

Fig. 91. Superior caval obstruction and pericardial effusion, anterior position. Injection was made into left antecubital vein. Note dilated left venous channels draining into a large vein coursing irregularly downward and to the left across the precordium (frames 1 and 2), filling of the heart from undiscernible entry site only after 9 seconds (compare frames 1 and 2), very late optimal visualization of the heart (center of frame 3), and void in activity around the heart due to effusion (frames 5 and 6).

and often by direct confirmation by visual inspection at the time of open-heart surgery. A large prospective clinical investigation to determine the sensitivity and accuracy of this method has not yet been carried out.

The results of intravenous radionuclide angiocardiography demonstrate that markedly abnormal and distinguishable scintiphotographic patterns are obtained in a variety of specific cardiovascular diseases. The immediate clinical relevance of the data gained combined with simplicity, speed, and safety of the procedure make this an attractive, useful, clinical procedure. Systems now available make it possible for the clinician to combine the qualitative aspects of the study with a variety of quantitative measurements. There is every reason to anticipate that the judicious application of the latter will eventually improve the sensitivity and diagnostic accuracy in many of the situations described in this chapter. However, it would be well to emphasize that a knowledge of the anatomic and gross physiologic derangements discernible on the qualitative study is a prerequisite to obtaining more refined and precise quantitative measurements.

In general, we recommend that radionuclide angiocardiography be performed as a screening procedure to select those patients requiring cardiac catheterization or contrast angiocardiography. Specifically, we suggest that the radionuclide procedure be considered under any of the following circumstances: (*1*) as a screening test in ambulatory or hospitalized patients with suspected congenital or acquired cardio-

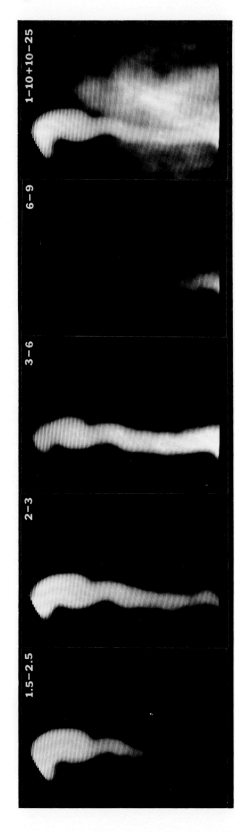

Fig. 92. Superior caval obstruction, anterior view. Injection was made into right antecubital vein. Note early transit of bolus through a large venous channel coursing downward across the precordium (frames 1–4) and late filling of the heart, which is poorly delineated due to bolus mixing and delay (frame 5).

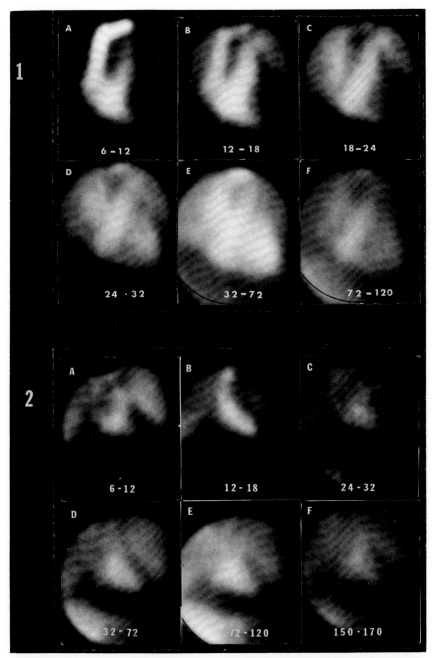

Fig. 93. Comparison of cardiac enlargement, and congestive heart failure (1, top panel), with pericardial effusion, left pleural effusion. (2, bottom panel). Anterior position, patient erect. In latter study note separation of heart from lung activity during first pass of bolus (frames A and B), appearance of saclike void in activity around the heart in later frames (C through F), and decreased activity at left lung base.

vascular disease; *(2)* in patients sensitive to radiographic contrast agents; *(3)* in patients too ill to undergo heart catheterization or contrast angiography; *(4)* serially, as a guide to the effectiveness of medical therapy; *(5)* serially, as a guide to the progression of disease; *(6)* preoperatively and postoperatively to document effects of certain specific operative interventions.

REFERENCES

1. F. J. Bonte, E. E. Christensen, and T. S. Curry, III: Pertechnetate ^{99m}Tc Angiocardiography in Diagnosis of Superior Mediastinal Masses and Pericardial Effusions, *Am. J. Roentgen.* **107:** 404–413, 1969.
2. F. H. Deland, and A. H. Felman: Pericardial Tumor Compared with Pericardial Effusion, *J. Nucl. Med.* **13:** 697–698, 1972.
3. L. P. Enright, N. E. Shumway, and J. P. Kriss: Acute Intracardiac Volume Changes Demonstrated with Radioisotopic Angiocardiography. *J. Thoracic Cardiovasc. Surg.* **64:** 136–141, 1972.
4. M. Ihnen, R. Teeslink, and J. L. Caldwell, Jr.: Evaluation of Lesions of the Aorta and Major Arteries by Scintillation Camera Technics, *Am. J. Surg.* **122:** 95–101, 1971.
5. G. T. Krishnamurthy, N. V. Srinivasan, W. H. Blahd: Pulmonary Hypertension in Acquired Valvular Cardiac Disease: Evaluation by a Scintillation Camera Technique, *J. Nucl. Med.* **13:** 604–611, 1972.
6. G. T. Krishnamurthy, M. A. Winston, E. R. Weiss, and W. H. Blahd: Demonstration of Collateral Pathways after Superior Vena Caval Obstruction with the Scintillation Camera, *J. Nucl. Med.* **12:** 189, 1971.
7. J. P. Kriss: Video Instrumentation for Radionuclide Angiocardiography, *Am. J. Cardiol.* **32:** 167–174, 1973.
8. J. P. Kriss, W. A. Bonner, and E. C. Levinthal: Variable Time-Lapse Videoscope: A Modification of the Scintillation Camera Designed for Rapid Flow Studies, *J. Nucl. Med.* **10:** 249–251, 1969.
9. J. P. Kriss, L. P. Enright, W. G. Hayden, L. Wexler, and N. E. Shumway: Radioisotopic Angiocardiography: Wide Scope of Applicability in Diagnosis and Evaluation of Therapy in Diseases of the Heart and Great Vessels, *Circulation* **43:** 792–809, 1971.
10. J. P. Kriss, G. S. Freedman, L. P. Enright, W. G. Hayden, L. Wexler, and N. E. Shumway: Radioisotopic Angiocardiography: Preoperative and Postoperative Evaluation of Patients with Diseases of the Heart and Aorta, *Radiol. Clin. N. Am.* **9:** 369–383, 1971.
11. D. T. Mason, W. L. Ashburn, J. C. Harbert, L. S. Cohen, and E. Braunwald: Rapid Sequential Visualization of the Heart and Great Vessels in Man Using the Wide-Field Anger Scintillation Camera, *Circulation* **39:** 19–28, 1969.
12. J. E. Morch, S. W. Klein, P. Richardson, G. Froggatt, L. Schwartz, and M. McLoughlin: Mitral Regurgitation Measured by Continuous Infusion of ^{133}Xenon, *Am. J. Cardiol.* **29:** 812–817, 1972.
13. D. Van Dyke, H. A. Anger, R. W. Sullivan, W. R. Vetter, Y. Yano, and H. G. Parker: Cardiac Evaluation from Radioisotope Dynamics, *J. Nucl. Med.* **13:** 585–592, 1972.
14. B. L. Zaret, H. W. Strauss, P. J. Hurley, T. K. Natarajan, and B. Pitt: A Noninvasive Scintiphotographic Method for Detecting Regional Ventricular Dysfunction in Man, *N. Eng. J. Med.,* **284:** 1165–1170, 1971.

PRINCIPAL EDITOR: RICHARD N. PIERSON, JR.

CONTRIBUTING EDITOR: DONALD C. VAN DYKE

5. Analysis of Left Ventricular Function

PHYSIOLOGICAL BACKGROUND

Most clinical morbidity in cardiology is associated with abnormalities of the left ventricle. Radiocardiographic techniques will achieve most utility to the cardiologist when he analyzes left ventricular function. Even when the mitral or aortic valve is the disease site, the most prominent pathophysiology is secondary to changes in left ventricular volume and contractility. The earliest and most subtle changes effect reduction in ventricular contractility and, therefore, the rate of volume change, or dV/dt, during systole. Later, reduced ejection fraction occurs. This phase is brief, since cardiac output must be maintained for the oxygen needs of the body; as ejection fraction falls, compensatory mechanisms result in increase in ventricular volume, such that a diminished ejection fraction (contractility) of a larger diastolic volume results in a normal stroke volume.

The most sensitive tests of ventricular function must be capable of appreciating changes in contractility which precede alteration of ejection fraction or diastolic volume. Indeed, the earliest evidence of left ventricular disease will not appear at rest, but only as a decreased functional reserve which is available to meet increased peripheral oxygen needs during exercise. To be most useful, a test of cardiac function should be sufficiently mobile to be applicable to exercising or postexercise subjects.

RADIOPHARMACEUTICAL AND INSTRUMENT REQUIREMENTS

The advent of technetium (^{99m}Tc) and other radionuclides which may safely be given in millicurie quantities as tracers for cardiopulmonary blood transit permit injection of a radionuclidic bolus emitting photons at the rate of 5×10^8/second.

With a detector viewing a small solid angle such as 5% (giving 95% loss of counts), and with additional losses due to collimator septal absorption and tissue absorption, precordial count rates in the order of 5×10^4/second are achieved. Other radionuclides have been and will continue to be evaluated for study of the pulsatile cardiac events (Chapter 9), but most work to date has been done with ^{99m}Tc compounds. Optimal collimator and crystal efficiency are essential for pulsatile flow measurements, and until now, the 140-keV gamma ray has been best matched by existing collimators and crystal design. Ventricular function may be studied either with the Anger-type camera, operated with the shortest possible dead-time (Chapter 8) to allow linear recording of count rates from 30,000 to 80,000 cps (counts per second), or with the inherently faster multicrystal camera. A single-probe method, having the advantage of portability and simplicity, may also be used with certain simplifying assumptions, as described in detail on page 136.

The frequency of cardiac contraction, 30 to 180/minute, requires a high-speed sampling capability in the range of 10 to 50 frames/second if the fine structure of the cardiac cycle, the left ventricular volume–time function (dV/dt), is to be studied. In radionuclidic studies, measurement of volume (V) is achieved via a measurement of radioactivity (A). Since a relationship exists between activity and volume (Chapter 7, p. 218), dV/dt may be estimated from the recorded parameter, dA/dt. Figure 94 shows the goal which is sought for isotope characterization of dV/dt. This filling and emptying function was derived from a high-speed contrast angiographic study using 64 frames/second (1). Rapid linear emptying during systole, followed by gradually accelerating diastolic filling, with a final "volume kick" following left atrial contraction, are well seen in this calculation of instantaneous ventricular volume from the geometrical method of area–length measurements. One criterion of success for any noninvasive measurement is ability to reproduce the pattern shown by this invasive and complex angiographic method.

TECHNIQUES

Two general methods have been used for analysis of scintillation camera cardiography. The first is based on rapid data sampling during the first pass of a radio-

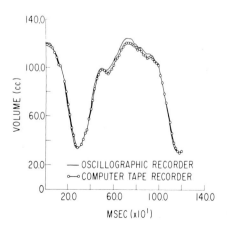

Fig. 94.

pharmaceutical after bolus injection. A complete high-frequency curve of left ventricular radioactivity, and hence a volume, is derived from "frames" processed at 10 to 50/second. The second method records two counts per cardiac cycle at preselected brief times at end systole and end diastole, starting several minutes after intravenous injection of ^{99m}TC-albumin, at a time when complete mixing of the radiopharmaceutical has occurred. The counts from 200 to 500 consecutive cardiac cycles over times of from 0.1 to 0.3 seconds in each cycle are integrated; thus this method requires much lower count rates to define ejection fraction. The basic assumptions required in Equation 1 are used in both methods.

$$\text{Ejection fraction} = \frac{\text{stroke volume}}{\text{end diastolic volume}}$$
$$EF = \frac{SV}{EDV} \tag{1}$$

and

$$\text{Stroke volume} = \text{end diastolic volume} - \text{end systolic volume}$$
$$SV = EDV\text{-}ESV \tag{2}$$

During the time of mitral valve closure, which is ventricular systole, the concentration of radionuclide remains constant in the left ventricle. During this period of fixed concentration, any change in counts is a linear function of the change in volume. With the opening of the mitral valve, a new concentration of tracer enters the ventricle (Fig. 95), higher if on the ascending limb of the precordial dilution curve, and successively lower as the bolus becomes increasingly diluted with untagged blood on the descending limb. During a typical left-heart transit from a right atrial injection, from 5 to 15 heart cycles will encompass the transit of the bolus, with much longer transit times in the case of small ejection fraction, large intervening blood volumes, or both. Typical transit activity functions are seen in Chapter 7, page 216. During systole dA/dt is precisely dV/dt, since tracer concentration in the ventricle does not vary. During diastole, however, the pattern of filling may be assessed from the shape of the dA/dt, but continuous change of tracer concentration in the ventricle during diastole prevents any direct estimation of instantaneous volume changes from the measured continuous changes in count rate. Explicit diastolic dV/dt can, however, be readily developed from the modeled reconstruction of the heart under study according to the techniques in Chapter 7. With appropriate selection of a region uniquely representing the left ventricle, the instantaneous dA/dt has been measured with analog (2) and digital (3) systems. The quality of the derived dV/dt curves is highly dependent on the actual count rate achieved, a dependence which is discussed in detail in Chapter 8. The recording of dV/dt must be made with a system providing a linear response to count rate. Analog systems must be well characterized for frequency response and for linearity. If a digital system is used, linearity is easier to check, but the choice of a time frame over which counts in the region of interest are integrated becomes critical: using frame rates in the range of 20 times the heart rate, systolic emptying and diastolic filling curves can be obtained which agree with those calculated from angiographic studies (compare Figs. 94 and 95). At frame rates in the range of 10 times or less

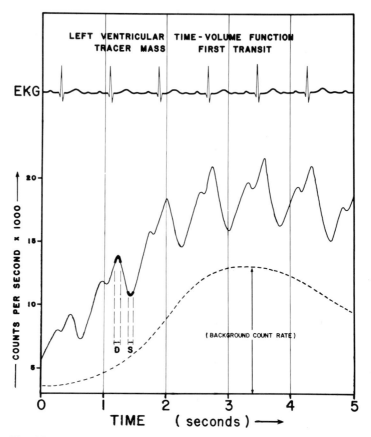

Fig. 95.

of the heart rate, the curves show increasing blunting of the dynamic changes involved. This effect is shown in Figure 96.

Electronic "Gating" Methods

In gating methods, the count rate from a carefully selected precordial view is recorded on an appropriate storage device, both during initial transit of the bolus and during a period of from 5 to 15 minutes after its equilibration in the blood volume. For analysis, the study is replayed and a region of interest is carefully selected from the first transit sequence to represent the left ventricle. The data from the later postequilibration period is then replayed from this selected region and the counts are recorded and integrated for selected time periods in the cardiac cycle to represent systole and diastole, defined from the simultaneously recorded electrocardiogram. The accuracy of these measurements is dependent on the spatial accuracy with which the left ventricle can be defined and the temporal accuracy with which end diastole and end systole are electronically selected by the gating process. Timing errors are likely to reduce the measured ejection fraction. In this technique, an

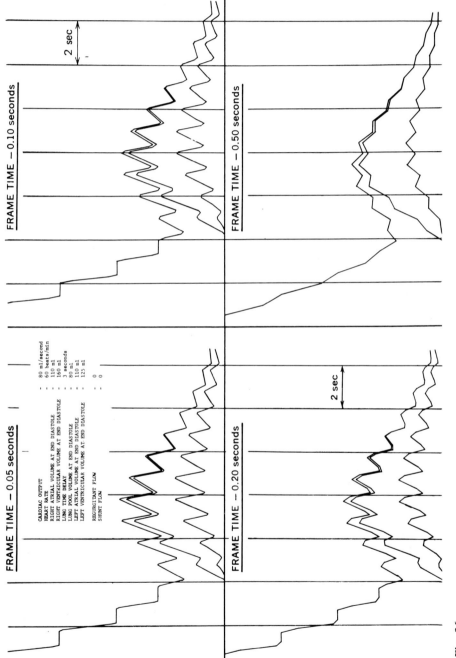

Fig. 96.

LAO view is usually used to select a region containing the left ventricle and a minimum of the left atrium and other sources of "background," "contaminating," and cross-talk counts. (Background, as used here, refers both to overlap with chambers and vessels other than the chamber of interest, and to scattered counts having origin elsewhere in the body or the environment. A rigorous conceptual handling of these two sources of background is presented in Chapter 7.) By integrating the counts during diastole for 300 cardiac cycles, Secker-Walker (4) was able to record approximately 50,000 diastolic counts over approximately 5 minutes of patient study time, with a dose of 10 mCi of ^{99m}Tc albumin. Using the same position, a net systolic count of approximately 30,000 was achieved, either by extending the patient study for an additional 5 minutes or by use of a computer which permits sampling and storing of both the systolic and diastolic counts from each cycle. By measuring both during the same time period, a changing physiological state is excluded as a source of error. Patient immobility during this period is critical to assure that the carefully selected region of interest continues to reflect activity uniquely in the left ventricle. The lower count rate during end systole is, of course, associated with a larger statistical deviation than that associated with end diastole. Table 2 indicates the relative merits of available methods for study of ejection fraction.

Correction for Background Counts

Figure 97 illustrates graphically the gating method of Secker-Walker and indicates one of the serious problems for which no rigorous solution has been established, although several approaches have been proposed. Photons emanating from a chamber under study are identified as "true counts" by appearance within a two-dimensional region of interest and by being within a predetermined energy range. Narrow-angle Compton scattering of photons in the body and in the collimator adds a significant number of counts which have origin outside the selected chamber. This background is low early in the bolus transit when the radiopharmaceutical is entirely within the right heart, but when the left ventricle is studied, the tracer has become widely distributed in the heart and chest, and background is on the order of 50% of total counts. If no subtraction for this background is made, the apparent ejection fraction measured is about half of that from near simultaneous angiographic studies (2, 5). For the case of first-transit studies, both explicit and empiric procedures have been suggested in the search for a solution to this problem. In one method, the time–activity functions are measured in neighboring chambers, an efficiency factor for the cross-talk counts from each of these areas into the region labeled "left ventricle" is applied, and a summed function is derived which represents the time-specific background. This approach is exemplified in the studies of Jones (3), in which the chamber-specific time–activity curves measured from adjacent regions are subtracted from the total left-heart curve. Approximately 80% of the background can be accounted for by this explicit means.

The empiric approach developed intuitively by Van Dyke and associates (2) and used by a number of other workers depends on the thesis that while the exact source of the background has not been specifically defined, there are counts recorded in the apparent two-dimensional image of the left ventricle which actually

Techniques

Table 2. Ejection Fraction by Radiocardiography

Parameters	First Transit	Gated for Counts (4)	Gated for Volumes (7)	By Single Probe
Patient position	LAO, 30°	LAO, 30°	RAO, 30°	LAO, 15°, or anterior
Tracer	^{99m}TcHSA ^{123}IHSA ^{99m}TcO$_4$	Either HSA	Either HSA	^{113m}In, ^{99m}TcHSA or ^{123}IHSA
Injection	Timed, into RA	I.V.; Timing not critical	I.V.; timing not critical	Superior vena cava or RA
Measurement	Counts	Counts	Volume from area–length	Counts
Limiting factor	Count rate	Patient immobility	Patient immobility	Probe placement
Assumptions	Correction for contaminating counts; accurate definition of region of interest	Region of interest: definition of aortic valve plane	Geometry of end systolic ventricle; Region of interest.	Contents of ring for background correction
Principle defect	Equipment expensive (computer)	Accurate gating essential	Depends on area–length formula for end systole	Uncertainty concerning accuracy of information.
Principle advantage	Most information including dV/dt, volumes of chambers, and spatial views	Least calculation required	Gives regional dysynergy information	Beside speed of performance and reporting
Equipment	Fast camera or mosaic system; special TV system or computer	Any camera and replay system; gating circuitry	Any camera and replay system; gating circuitry	Probe, ratemeter, recorder; simplest and least expensive

have origin in regions in front of, behind, and around this image. Although the source of these counts is not precisely known, the background can be simultaneously measured in an area as near as possible to the actual image; whatever background affects the apparent "left ventricle" must also affect a ring drawn as closely around it as possible. Figures 98a–98e show the effects of changes in choosing the limits of the left ventricle according to Van Dyke's method. The first panel (Fig. 98b) shows the count maps of the left ventricle in such a way that the X and Y (dimensions in a plane) and Z (count rate) axes are displayed; subsequent Figs. (98c–98e) show the time–activity functions at 0.1-second intervals. It is clear from this study that defining the diastolic ventricle by too small a region of interest results in a damped recording of left ventricular count rate changes and in too small a calculated ejection fraction. In a similar manner, it may be shown that correct

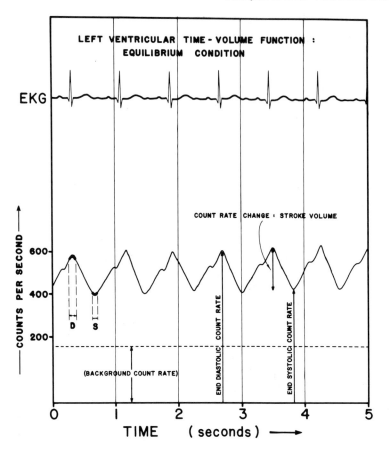

Fig. 97.

outlining of the left ventricle, but too narrow a ring for background subtraction, results in a small subtraction and, therefore, a too-low ejection fraction. Too wide a ring has the opposite effect. A selection of the ring width such that the amount of estimated background subtracted results in a constant ejection fraction provides a rational correction, which is, however, subject to the assumption that ejection fraction in fact remains constant while the first transit of the tracer bolus is occurring. This assumption is probably tenable within a 20% error in patients with sinus rhythm and without unusual degrees of cyclic change in intrathoracic pressure affecting venous return. These techniques have been developed for analysis of first-transit data, during the collection of which the operator has the advantage of being able to use temporal separation of the chambers to define their spatial limits without interfering counts from adjacent chambers. The cross-talk contributions can be discretely mapped individually for each patient during this first transit, when the counts over the left ventricle, monitored during right ventricular transit prior to arrival of the tracer bolus in the left side of the heart, provide a specific correction. Measurement of the left-atrium to left-ventricle cross-talk is much more difficult because of the temporal and spatial propinquity of these chambers; this is quantitatively the most important correction. Data for this "efficiency factor" can be

Techniques

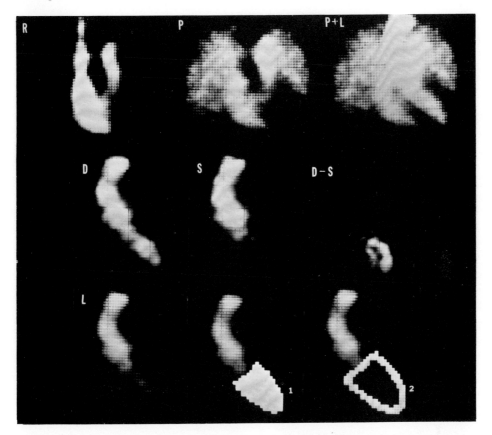

Fig. 98a. Radioisotopic angiocardiogram, anterior position following intravenous injection of 10 mCi ^{99m}Tc-albumin to a patient with valvular aortic stenosis; computer display on 64 × 64 matrix. Top row: R, right-sided phase; interventricular septum is displaced to the right. P, pulmonary phase; there is residual activity in the right heart, but the left ventricle is not yet filled. P + L, pulmonary + left-sided phase (note the LV cavity occupies only the middle portion of the LV region; the void in activity separating LV from RV and L lung represents the thickness of myocardial wall). Middle row: D, summed end-diastolic images obtained during fast-framing of study at 10 frames/second during first pass of bolus. S, summed end-systolic images. D-S summed diastole minus summed systole (i.e., image of ejected blood); the relatively diminished activity at the inferior portion of the LV may be caused by hypertrophied papillary muscles. Bottom row: L, average left-sided image, area *1*, area of interest for LV (see explanation for size and shape of this area in Fig. 2), area *2*, "ring" background for LV.

gained from the occasional patient in whom left atrial and left ventricular injections have been carried out during catheterization, but the general applicability of this correction is doubtful, since small changes in the positioning of the region of interest may result in marked alterations in the volume of the left atrium which is directly in the region designated as left ventricle. Further studies with test systems and with patients are currently in progress to define more precisely the contribution of counts emanating from the left atrial chamber to the planar region designated as left ventricle.

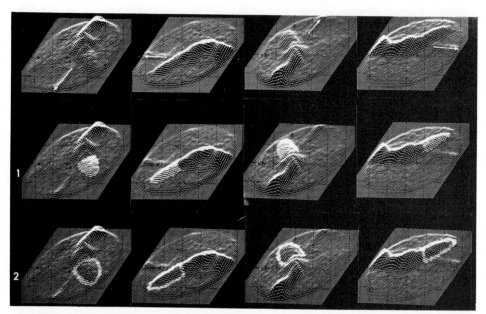

Fig. 98b. Isometric displays of left-sided phase, same study as in Figure 98a, Top row: at left contains a conventional view with arrow over abdominal aorta pointing to direction of blood flow, and in addition a series of frames show successive clockwise rotation of 90°. Middle row: same views as at top, but with area *1* superimposed; note how the area must be as large as it is to include all activity (in fact, it should be somewhat larger along right border of the LV). Bottom row: same views with superimposed "ring" correction (note "ring" lies outside the mound of LV activity).

It is reasonable to apply these background correction factors derived from first transit studies to equilibrium-time measurements involving gated image integration. For the greatest accuracy obtainable, each patient should have a set of correction factors measured. If appreciation of serial changes in ejection fraction are required, a standard set of factors, or no factor at all, may be applied. However, if terms such as ejection fraction, which have general meaning in cardiology, are to be used in radiocardiography, it will be essential to the credibility of the method that the ranges of normal be established for this technique and that this range be comparable to that well established in the cardiologist's lexicon.

Spatial Volume Methods

Ashburn (6) and Strauss (7) have used a different method of calculating end diastolic volume and ejection fraction, based on gating for integrated images, and angiographic techniques of analysis of these images for area–volume measurements. In this method gating is used to define end-systole and end-diastole in a manner conceptually identical to that described for the counts per volume method. However, the planar dimensions of the left ventricle in the right anterior oblique projection are viewed, and the well validated angiographic method (8, 9) of estimating the ventricular volume from the major and minor axes in this plane is applied. In this approach, the ventricle is assumed to be an ellipsoid of revolution. A perfectly

Techniques

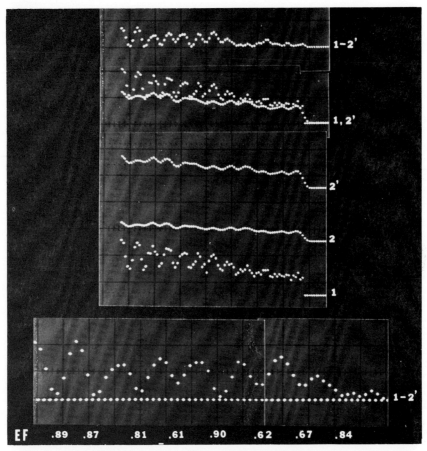

Fig. 98c. Calculation of ejection fraction from high-frequency data display during first pass of radiopharmaceutical; same case as in Figure 98a.

    Graph *1*: area *1* of LV, raw data, 10 frames/second
    Graph *2*: area *2* (5-point smooth), 10 frames/second
    Graph *2′*: area *2* multipled by a numerical factor which results in approximately equal value for 1 and 2′ after the bolus has passed.
    Graphs *1,2′*: superimposed graphs *1* and *2′*
    Graph *1−2′*: graph *1* minus adjusted ring correction *2′*

At bottom (*1−2′*) is an expanded version of the final graph at top showing the corrected intraventricular activity with each heart beat. Ejection fraction (EF) is calculated from the measurement of ejected activity and total activity for each beat read from the graph. Note that there is moderate beat to beat variation (mean value = 0.78).

orthagonal view of the ventricle is assumed. Small errors in this assumption would result in very slight foreshortening and would not materially affect the calculated volumes. The right and left ventricular cavities are seen overlapping in this view, but the left ventricle forms the outermost boundary in the great majority of instances, and it is the left ventricular measurements which are recorded. Definition of the aortic valve plane is critical to this approach; this is usually done by careful viewing of the first-transit images, in which the valve plane can usually be well

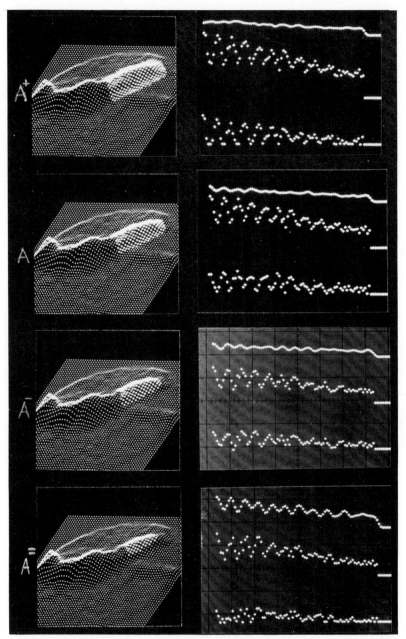

Fig. 98d. Study showing effect of variation of size of area of interest on calculations of ejection fraction using method depicted in Figure 3, same case as in Figure 98a A⁺, area of interest 2 pixcel widths too large; A, ideal placement of area of interest: A⁻, area of interest 2 pixcel widths too small, A⁼ area of interest 4 pixcel widths too small. To the right of each image, reading from top to bottom, are (*a*) activity of the ring correction of 2 pixcel widths (5 point-smooth), (*b*) activity of the designated LV area, and (*c*) corrected LV curve (see enlargements Fig. 98e).

Techniques

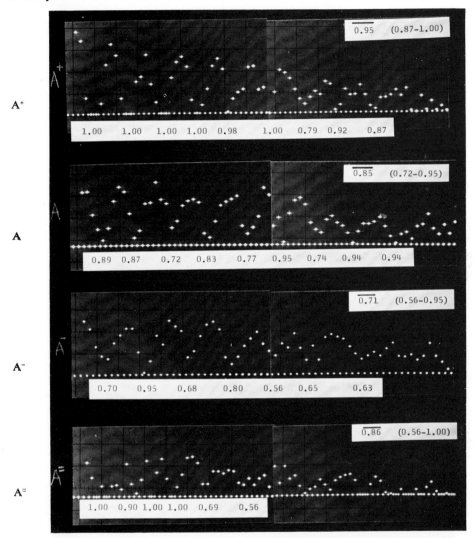

Fig. 98e. Corrected curves with calculations of beat by beat ejection fraction for the areas of interest described in Figure 98d. Ideal placement (A) shows mean ejection fraction of 0.85 with the smallest range (0.72–0.95). Making area of interest too large (A$^+$) yields values for ejection fraction which are obviously too high. If area of interest is too small (A$^-$), value for EF is either too low with greater beat-to-beat variability or curve becomes largely uninterpretable (A$^=$).

seen, and in which the assumption that the left ventricle forms the outer cardiac boundary in the view selected can be verified.

Data rates and frame times; in search of optimal solutions

Van Dyke (2) has published curves produced at 0.03 second/frame, the natural refresh rate of a television system. This produces a smooth volume–time function, as would be expected from the short frame time. However, for any given count rate, shorter framing times will give a lower number of counts per frame, with concurrent increase in the statistical error of the measurement. The optimal frame length is an inverse function of both heart rate and count rate. Count rate within the region of interest (ROI) in the left ventricle is in turn a function of initial dose, method of injection, and the ratio of cardiac output to central blood volume. All existing methods for estimating dV/dt operate close to the lower limits of statistically acceptable count rates (100–200) counts/ROI/frame) for heart rates between 50 and 150. It would therefore be desirable to be able to set the frame time retrospectively by a two-variable minimization formulation after the study has been completed, dependent on heart rate and data density (count rate) in the area of interest. Optimal frame time is highly dependent on statistical constraints, and the nature of the minimization program requires a very substantial amount of computation, such that it can only be handled within reasonable analysis times by a high-speed digital computer. Given this method for data collection and analysis, the incoming count data stream should be collected in list mode for subsequent framing based on the nature of the data. Chapter 8 deals explicitly with definitions, techniques, and strategies of data recording and manipulation.

THE SINGLE-PROBE STUDY

Measurements of heart function with a single response probe were initiated in 1927 when the first photon-sensitive instrument, a cloud chamber, was used to measure and record the transit of a tracer through the heart (10). Lammerant (1956) (11), MacIntyre (12), Huff (1958), Donato (1962), Folse (1963), and many others were involved in the development of this technique, which was attractively simple to perform, if not to interpret, and depended on inexpensive and widely available instrumentation. Several factors inhibited its use and wide acceptance during that era.

First, short-lived radionuclides were not available; because high dosage of longer-lived radionuclides like Iodine-131 in the form of ^{131}I-human serum albumin would have been associated with an unacceptably high radiation exposure to the patient, relatively small doses had to be used, and the data density was accordingly low. (Introduction of ^{131}Iodipamide shortened the effective patient exposure time by a factor of 10, but its use was never widespread.) With doses of from 10 to 60 μCi external precordial count rates in the range of 1000 cps at peak, and 150 cps at equilibrium were achieved, but the curves required smoothing by ratemeter time constants of 1 to 2 seconds, enough to distort the shape of the transit curve significantly. However, the measurements did permit calculation of the cardiac output. The substitution of ^{125}I for ^{131}I by Gorten (16) reduced patient dose by a factor of 10, but high tissue absorption of the low-energy (25- and 35-keV) photons in the heart and chest wall resulted in great sensitivity to minor changes in superficial and irrelevant blood pools, such as the internal mammary artery, and concomitantly low counting efficiency for deeply situated blood volumes, such as the ventricles.

Second, available analog ratemeters, if operated at long time constants, added a delay to curve transcription which did not affect cardiac output calculation, but

The Single-Probe Study

markedly affected mean transit times and derived volumes. Digital ratemeters and quick-response analog ratemeters, which overcame the undesired smoothing effect of older analog ratemeters, were not widely available.

Chiefly for these reasons, single-probe studies never achieved currency, although reviews in 1962 (14), 1966 (5), and 1973 (16) reiterated the soundness of the mathematical principles and stressed the richness of information available from a fully analyzed indicator-dilution curve.

In 1972, following awareness of the depth of information available from region-of-interest gamma-camera studies of the heart in dynamic mode (2,3), the single-probe method was reintroduced (17) under conditions utilizing the lessons learned from such camera studies and taking advantage of the availibility of ^{113m}In. Several important refinements of interpretation were made possible by employing millicurie doses and using a high-frequency recording system. ^{113m}Indium is a radiopharmaceutical with ideal characteristics of availability (generator), effective half-life (100 minutes), protein tagging (to transferrin, if the generator eluate is injected undiluted directly into the blood stream), and a photon energy of 393 keV (which assures near isoefficiency of counting regardless of tissue depth). The absence of beta radiation is an additional important factor in reducing patient radiation exposure to 16 mrad/mCi administered. Standard equipment is used: a 2-in. NaI crystal housed in a lead collimator, a fast response ratemeter, and a strip chart recorder. Total expense for this equipment is less than $10,000, and it is often already available in a Nuclear Medicine Department. The probe–recorder assembly is readily portable, and it has been used to study acutely ill patients at the bedside, providing quantitative measurements of cardiac output, pulmonary blood volume, and ejection fraction within the hour, and preliminary useful qualitative information within seconds after injection (17–19).

Cardiac output is calculated after measuring the area under the curve inscribed by the strip chart recorder during the first transit of injected bolus (area *1*), the area under a 1-minute recording taken 5 minutes later when the tracer dose has equilibrated in the blood volume of the patient (area *2*), and the radioactivity of a sample of blood at the 5-minute interval. The method requires assumption of uniform mixing of the label in the blood. Since recirculation, or "second transit," occurs before the first transit curve has reached zero, the well-recognized forward extrapolation of the exponential portion of the washout curve from the left heart is used to define the limit of this area. Integration of these areas may be accomplished by means of a planimeter or by digital summing. This use of the Stewart–Hamilton approximation is widely published and accepted (20); an explicit formulation for use when a radioisotope rather than an intravascular dye is used is contained in Chapter 7. The calculation of cardiac output (CO) is as follows:

$$\frac{\text{area } 2}{\text{area } 1} \times \text{blood volume} = \text{CO}$$

(The dimensions of this formula are $\frac{\text{cm}^2/\text{minute}}{\text{cm}^2} \times \text{liters} = \text{liters/min.}$)

With the data recorded at a time constant of 0.05 seconds or less a pattern is obtained which is exactly analagous to that from the curves derived from the area-of-interest for the left ventricle in high-frequency (i.e., fast-framing) scintillation-camera studies.

To measure cardiac output, placement of the probe is not critical providing it is viewing the left heart predominantly. For measurement of ejection fraction however, it is critical that only the left ventricle be viewed (or that the activity of the left ventricle eventually be distinguishable from activity arising elsewhere) for the reasons presented on pages 128–132. All the left ventricle must be seen. As much as possible of all other chambers and organs must be excluded, and the contribution for regions other than the left ventricle must be estimated, since these form the "contaminating-count" contribution.

With a radiopaque marker on the chest wall over the presumed location of the left ventricle, a portable supine antero-posterior chest x-ray is taken in midinspiration and in the same position in which the radionuclide study will be made. The distance and direction of the marker from the center of the left ventricle is noted, and an appropriate correction for probe placement relative to the marker is made. The correct position for the probe is then drawn on the chest with a felt pen to facilitate accurate repositioning for serial studies.

A standard scintillation probe with a 2 in. × 2 in. sodium iodide crystal is used. A portable stand carries the probe, the high-voltage power supply, count ratemeter (with 10 million counts/minute capacity and time constant of less than 0.1 second), and a high-speed strip chart recorder (at least 30 cm/minute). Linearity of the entire system may be demonstrated by recording activity while rotating over a 1-mCi ^{113m}In source an absorber wheel with 10 segments giving from 0 to 100% absorption.

Collimation for obtaining the LV radioisotope count-rate curve is as shown at the top of Figure 99. A $1\frac{3}{8}$-in. circular port (in $\frac{1}{2}$-in. lead collimator) is positioned over the midpoint of the LV. A bolus of 1 mCi ^{113m}In is flushed through a central venous catheter and the activity over the LV is recorded with the probe perpendicular to the chest wall. A recording of residual activity is made after 5 minutes of equilibration to obtain the value after mixing. A blood sample is taken at this time for determination of blood volume. From these measurements cardiac output (CO) is determined, as explained earlier.

In order to define the contaminating counts in the field of view, collimation of the probe is then modified to the configuration shown at the bottom of Figure 99. The LV is eclipsed by a $2\frac{3}{16}$-in. circular shield placed $1\frac{3}{8}$-in. out from the probe face ($4\frac{3}{8}$-in. from the sodium iodide crystal) and maintained in position by a styrofoam cone. The eclipsing disc is placed over the midpoint of the LV to absorb counts from the LV while viewing a ring around the LV which is representative of the sources of contaminating counts. For this recording a second bolus of 500 μCi ^{113m}In is given. The two records are matched temporally by aligning the times of entry in the right heart and also aligning the amplitude (see below) to provide the correct base line from which ejection fraction (EF) is calculated.

Initial studies using a scintillation camera with data storage and area-of-interest capability showed that the ejection fraction, read from recording of the left ventricular area of interest without modification of base line, was approximately one half the true value. Dynamic studies of a single-chamber model demonstrated that this discrepancy was not due to a fault of the recording system and must, therefore, be due to activity in adjacent structures. The true base line must be one which results in a calculated relatively constant ejection fraction throughout the left ventricular phase of bolus transit. This requirement is met by any one of a family of

The Single-Probe Study

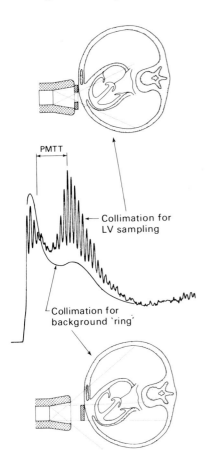

Fig. 99. Schematic representation of single-probe collimator configuration for recording ventricular radioisotope concentration curve and background "ring." The pattern shown (center) is characteristic of those obtained in a normal subject.

base-line curves (Figure 100). Using the results of contrast angiography as the standard, it was apparent that the correct base line was one which intercepted the left ventricular curve at its nadir after passage of the bolus and before recirculation. In practice, this is determined by recording the ring-shaped area of interest around the left ventricle at a sensitivity level which brings the post-transit low point to the same value as the low point following left ventricular washout. The optimum width of the ring was determined by trial and error in direct comparison to ejection fraction determined by contrast angiography (2).

Ejection fraction is calculated beat by beat during left ventricular washout by subtracting the background value from each pair of values representing end diastole and end systole. Left ventricular end diastolic volume is obtained by dividing the stroke volume, as obtained from the cardiac output and heart rate, by the ejection fraction.

The size of the pulmonary compartment is calculated from the fundamental relationship that the mean transit time is the ratio of a volume to the flow through it. Knowing flow (cardiac output), one need only determine the correct pulmonary transit time in order to calculate pulmonary blood volume. Thus, in the precordial probe method the term pulmonary blood volume (PBV) refers to the volume of blood intervening between the right and the left heart, including the pulmonary

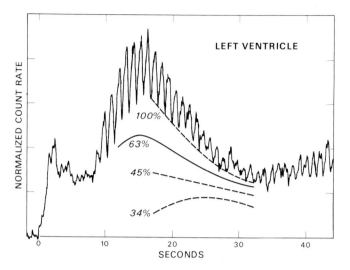

Fig. 100. Recording from left ventricle showing some of "family" of curves which result in constant ejection fraction through left ventricular phase. Solid curve (63%) represents correct base line obtained by simutaneous quantitative angiocardiography. (From *J. Nucl. Med.* **9,** by permission.)

arterial, capillary, and pulmonary venous circulation. Determination of the pulmonary transit time (PTT) from precordial radionuclide activity curves still presents problems. Such measurement has been the subject of a number of excellent theoretical and experimental studies, leading to the mean-transit-time, the left-peak, and the simulation methods (21). When applied to relatively normal subjects or to patients with only moderately compromised pulmonary vasculature, these methods have all been shown to yield comparable results which are in agreement with results obtained by double-injection indicator-dilution methods. However, when the bedside technique is applied to critically sick patients with cardiopulmonary collapse, one frequently finds the right ventricular washout so slow (because of RV dilatation and small stroke volume) that either the mean-transit-time or left-peak method gives a negative value for PBV (17–19). In such patients one is forced to something approaching the old peak-to-peak method of Eich (20), to use of the mean transit times, or to a computer simulation method (22).

In patients with a markedly dilated right ventricle or with a large part of the pulmonary circulation eliminated (as in massive pulmonary embolus) or both, use of the method of Giuntini and co-workers (14) for calculation of PBV (time when 37% of RV peak count is reached is used as the mean time of exit from the RV) results in measurements of RV mean exit time that occur at the same time or even after the time the counting peak in the LV is observed.

The use of peak-to-peak time, on the other hand, results in large values for pulmonary blood volume when the latter is known from other indications to be normal or decreased. A reasonable minimum value (100–200 ml) for PBV was obtained in such cases only by using the time on the washout downslope at which 70 or 80% of the RV peak was reached, rather than the 37% of Giuntini (14). Van

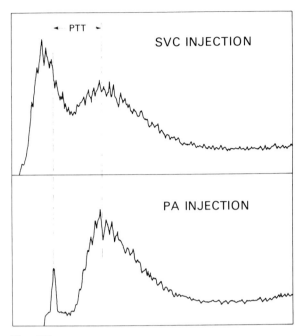

Fig. 101.

Dyke and his colleagues currently use the time on the washout slope when concentration has fallen to 75% of peak. This choice results in frequent overestimates of pulmonary blood volume, in order to report consistent values and to avoid the absurdly low and even negative values obtained with the 37% method. Figure 101 illustrates how use of this time intercept as the initial time and use of the left ventricular peak as the terminal time compares with the results obtained by introduction of the label into the pulmonary artery. The curves shown were obtained by first introducing the label into a Swan–Ganz flow-directed catheter in the superior vena cava and then advancing the catheter to the pulmonary artery for a second injection. This and other similar practical demonstrations (11, 12) tend to validate this empiric approach.

Computer simulation has made it clear that the use of LV peak time is a poor choice of end point for the calculation of PBV (13), but has not suggested any better single end point applicable to the bedside method, without the use of mathematically rigorous methods, such as mean transit time analysis.

A typical single-probe precordial radiocardiogram from a normal subject is shown in Figure 102. The normal radiocardiogram consists of right- and left-heart peaks with an intervening "pulmonary valley." The various diagnostic features of such a record, both contributing to pattern recognition and providing the basis for quantitation, are listed in the figure caption.

Van Dyke and his colleagues at the Stanford Medical Center have compared ejection fraction and cardiac output obtained with this technique with those from contrast angiography and Fick methods in 14 patients. The angiographic ventricular volumes were calculated from single-plane RAO ventriculograms using the area–length method. In patients with ejection fraction ranging from 9 to 79%, the

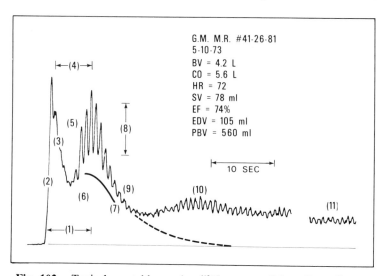

Fig. 102. Typical portable probe ^{113m}In precordial radiocardiogram from a normal subject. The essential features of such a strip chart recording, which is immediately available at the patient's bedside, are: (1) overall transit time through the central circulation (time from injection to peak of LV) prolonged in circulatory failure and in the presence of increased PBV; (2) rate of appearance of label in RV (flow from SVC to RV) is prolonged in severe circulatory failure; (3) downslope of RV washout curve: slow in circulatory failure and in the presence of dilated right heart chambers; (4) pulmonary transit time (PTT) decreased with decreased pulmonary blood volume and viceversa, and increased in circulatory failure; (5) "pulmonary valley" between RV and LV: obliterated with markedly decreased PBV or markedly dilated right heart chambers. (6) Area under the first pass of the indicator: when related to the concentration after mixing in the blood, 11, is inversely related to cardiac output. (7) Counts from surrounding tissues after eclipsing the LV: provides the baseline against which EF is measured. (8) Ejection fraction (EF): related to 7. (9) Washout slope of LV: slow in failure and in left-to-right shunt. (10) First recirculation: shortened in AV fistula, blunted in circulatory failure, and obliterated in left-to-right shunt. (11) Concentration after complete mixing: small (as compared to area under first pass, 6) in low cardiac output or exceptionally large BV. (From H. Parker & D. C. Van Dyke et al, in *Diagnostic Nuclear Cardiology;* C. V. Mosby Co., St. Louis, Mo., 1974, by permission.)

correlation coefficient for the two methods was .91 ($p < .01$). Comparison of cardiac output values yielded a correlation coefficient of .88. These correlations are in good agreement with those found by the group working in Denver (11,13). We conclude that these results validate the usefulness of this noninvasive technique for the bedside measurements of ejection fraction and cardiac output.

Large changes in ejection fraction, ventricular diastolic volume, and pulmonary blood volume are the rule in life-threatening crises such as major pulmonary embolism, right ventricular dysfunction (cor pulmonale), and left ventricular dysfunction. When one of these or some other cardiovascular emergency is superimposed

The Single-Probe Study

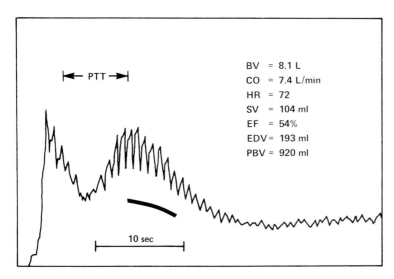

Fig. 103. Radiocardiogram of a patient with 3⁺ mitral regurgitation. In the presence of regurgitation, EF includes reverse flow so that calculation of EDV from EF underestimates the true volume by the regurgitant fraction.

on severe lung disease, the relative contributions of lung and heart disease to morbidity may be ill-defined on clinical grounds and yet may be clearly delineated by radiocardiography. In such circumstances, the single-probe method has been of proved value.

Clinical Studies

The following patients present the major characteristic patterns associated with right-heart, pulmonary, left-heart, and combined dysfunctions.

Figure 103 shows the precordial single-probe radiocardiogram from a 59-year-old man referred for evaluation of nocturnal angina of rapidly increasing frequency and severity, and the recent development of a murmur of mitral regurgitation. At coronary angiography he was found to have significant lesions in both the posterior descending and right coronary arteries, and 3+ mitral regurgitation. Hypokinesis of the inferior wall of the left ventricle was noted at rest. At catheterization his left ventricular end diastolic pressure (LVEDP) rose with exercise from 12 to 18 mm Hg. At surgery the left atrium and left ventricle were moderately enlarged and there was rupture of the chordae tendinae to the posterior leaflet of the mitral valve. His radiocardiogram (Fig. 103) showed abnormally slow washout from the right ventricle and a normal pulmonary transit time and pulmonary valley followed by a grossly normal left ventricular peak. Calculation of left ventricular end diastolic volume by this method requires that no valvular insufficiency be present (mitral or aortic). For the calculation, the assumption is made that the ejection fraction represents forward flow, which is stroke volume. In the presence of regurgitation, ejection fraction includes reverse flow and the calculated diastolic volume is erroneously small. Thus, the value of 193 ml for end diastolic volume in this patient is an underestimate.

I.R. was a 49-year-old alcoholic female admitted in pulmonary edema following a 3-month history of progressive dyspnea on exertion and fatigue. After resolution of the pulmonary edema, cardiac catheterization was performed. In addition to generalized cardiomegaly there was poor ventricular contractility. Her left ventricular end diastolic pressure rose to 20 mmHg with exercise. The final clinical diagnosis was cardiomyopathy of unknown etiology. The radiocardiogram (Fig. 104) was done at the time of catheterization and illustrates two important features. In the LV area of interest recording the RV peak is reduced to a shoulder. This occurs when the LV is sufficiently enlarged so that positioning over its midpoint results in including little of the RV. The RV peak included in the tracing (upper partial curve) is from the background recording which includes more of the RV. Note that the LVEF is pathologically small (26%) and that when related to the SV (53 ml) indicates a dilated LV (204 ml). Thus her radiocardiogram is typical of moderate, isolated left ventricular dysfunction manifested by dilatation and reduced ejection fraction.

F. L. was a 46-year-old man who presented with severe congestive heart failure of 6-weeks duration. Chest x-ray showed cardiomegaly with biventricular enlargement. At cardiac catheterization poor contractility of both ventricles was noted. The mean PA pressure was 32 mmHg, pulmonary wedge pressure 17 mmHg, LV end diastolic pressure 17 mmHg, and cardiac output 1.9 liters/minute. The final clinical diagnosis was progressive cardiomyopathy of unknown etiology. His radiocardiogram (Fig. 105) showed marked prolongation of overall transit time, associated with low ejection fraction and reduced cardiac output. This pattern is typical of biventricular failure.

C. D. was a 55-year-old woman who was admitted with a 4-year history of orthopnea, palpitations, and shortness of breath. During the past 6 months she had developed cough, hemoptysis, and pleuritic chest pain. Recent electrocardiograms had shown progressive right ventricular hypertrophy. Serial chest x-rays had shown

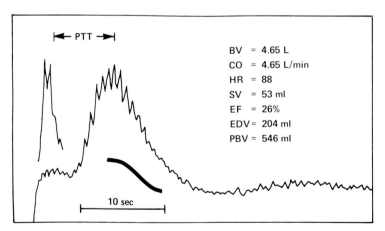

Fig. 104. Radiocardiogram from a patient with cardiomyopathy of unknown etiology. The radiocardiogram is typical of moderate, isolated LV dysfunction consisting of dilatation and reduced EF.

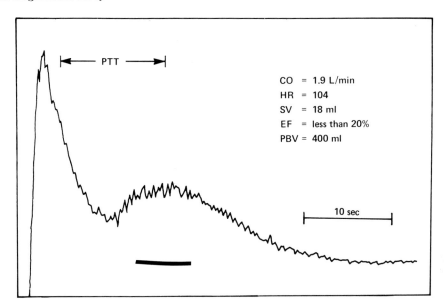

Fig. 105. Radiocardiogram from a patient with progressive cardiomyopathy. The radiocardiogram shows marked prolongation of overall transit time, associated with low ejection fraction and reduced cardiac output.

progressive enlargement of the main pulmonary arteries. Right-heart catheterization at the time of angiography revealed a right ventricular pressure of 88/12 mmHg and a pulmonary artery pressure of 94/48 mmHg, with a mean pressure of 62 mmHg. There was marked enlargement of the right ventricle and the main pulmonary arteries, but the peripheral pulmonary arteries appeared essentially normal. This roentgenographic finding can be seen in primary pulmonary hypertension. Her radiocardiogram (Fig. 106) was grossly abnormal with loss of the pulmonary valley, immersion of the LV peak in the RV washout pattern, but maintenance of the pulmonary transit time. This is the characteristic pattern of isolated right ventricular dysfunction (marked dilatation with slow washout of the tracer). Note that the values derived for pulmonary blood volume, ejection fraction, and end diastolic volume were within normal limits.

B. Y. was a 75-year-old man who collapsed 2 days following cystectomy and ileal loop diversion for bladder carcinoma. Arterial blood gases showed a pO_2 of 73 mmHg and pCO_2 of 26 mmHg. Ventilation–perfusion lung scintigraphy revealed multiple perfusion defects in regions with normal ventilation. His bedside radiocardiogram (Fig. 107) was grossly abnormal with greatly diminished pulmonary valley and rapid transit of the label from RV to LV (decreased pulmonary transit time.) Washout from the RV was not unduly prolonged, indicating that loss of the pulmonary valley was primarily due to loss of pulmonary blood volume. This pattern of marked volume reduction is characteristic of massive embolus. Note the respiratory fluctuations in the LV portion of the record and the moderately reduced ejection fraction, indicating some degree of LV dysfunction. The diagnosis of pulmonary embolic disease was further substantiated with pulmonary angiography.

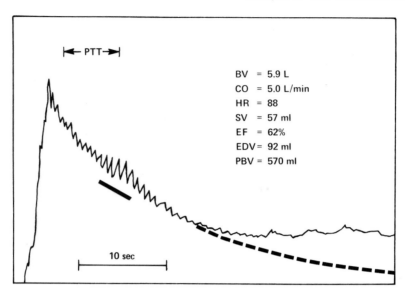

Fig. 106. Pattern from a patient with pulmonary hypertension and cor pulmonale. The radiocardiogram is grossly abnormal with loss of the pulmonary valley and immersion of the LV peak in the slow RV washout pattern, but maintenance of the PTT. This pattern is characteristic of right ventricular dysfunction.

W. S. was a 33-year-old female admitted with a 1 week history of increasing shortness of breath and pleuritic chest pain, and a 1-day history of hemoptysis. On admission she was found to be in mild respiratory distress; good breath sounds were noted over both lungs. On room air arterial pO_2 was 67 mmHg and pCO_2 was 34 mmHg, with pH of 7.43. Chest x-ray showed a left pleural effusion, retrocardiac infiltrates, and basilar atelectasis. The EKG was normal with a sinus rate of 100/minute and an axis of 0°. A perfusion lung scan revealed complete absence of perfusion to the left lung (Fig. 108,*A*). A pulmonary arteriogram on the day of admission (Fig. 108,*B*) showed total occlusion of the left main pulmonary artery, a pulmonary artery systolic pressure of 40 mmHg, and mean PA pressure of 25 mmHg. She was subsequently anticoagulated, and during the following 2 weeks she improved clinically. However, a repeat lung scan revealed that there was no resolution of the left pulmonary artery thrombus during her 2-week hospitalization. Congenital agenesis of the left pulmonary artery was excluded by the finding of normal left-lung vasculature on previous chest x-rays. Her bedside radiocardiogram was grossly abnormal (Fig. 109), with greatly diminished pulmonary valley and markedly shortened pulmonary transit time. This, like the previous pattern (Fig. 107), is characteristic of markedly reduced pulmonary blood volume and points to the diagnosis of massive pulmonary embolus. Note the abnormally small LV ejection fraction of 30%.

G. B. was a 35-year-old man with known idiopathic pulmonary hypertension. Extensive evaluations during the past 5 years had not suggested a possible etiology, and he had failed to improve on either anticoagulation or corticosteroids. His chest x-ray showed massive dilatation of the pulmonary arterial tree and moderate RV enlargement. His most recent cardiac catheterization 4 years previously revealed a

The Single-Probe Study

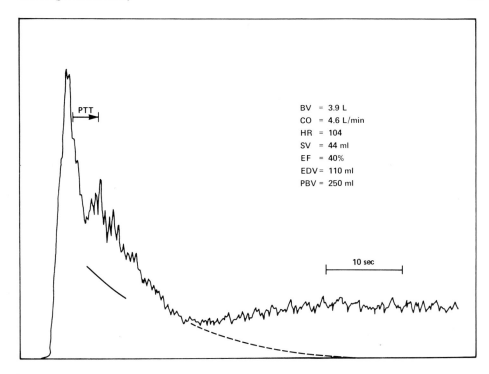

Fig. 107. Radiocardiogram from a patient with recent, multiple, large pulmonary emboli. The radiocardiogram is grossly abnormal with greatly diminished pulmonary valley and rapid transit of the label from RV to LV (decreased PTT). This pattern is characteristic of marked reduction in pulmonary vasculature. Note the respiratory fluctuations superimposed on the LV portion of the record.

mean pulmonary artery pressure of 45 mmHg with diminished visualization of smaller vessels on angiography. During a routine clinic visit, a radiocardiogram was performed (Fig. 110) that was grossly abnormal, showing loss of the pulmonary valley, slow passage of tracer through the RV phase, and displacement of the LV phase to the right (prolonged pulmonary transit time). Slow washout from the RV indicates dilatation of the right heart, and the late appearance and small amplitude of the LV curve reflects an abnormally large volume between right and left heart, or increased pulmonary blood volume. The calculated PBV was twice normal, 1300 ml.

E. B. was a 49-year-old woman who was referred for evaluation of a heart murmur that was first noted during a routine physical examination. The chest x-ray showed evidence of pulmonary hypervascularity. Cardiac catheterization showed an atrial septal defect with with a 3:1 pulmonary/systemic flow ratio. Her pulmonary artery pressures were 38 mmHg systolic and 21 mmHg mean at rest, and with exercise 47 mmHg systolic and 33 mmHg mean. Wedge pressures were normal both at rest and with exercise. Her radiocardiogram (Fig. 111) was grossly abnormal because of apparent delayed washout in the LV phase of the record. The pattern is characteristic of left-to-right shunt, that is, early reappearance of the tracer bolus in the LV. Reappearance of the label in the RV is obscured in the

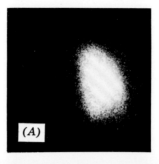

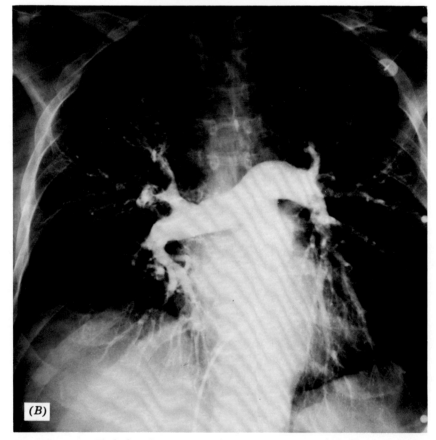

Fig. 108. (*A*) Perfusion lung scan (above) and (*B*) pulmonary arteriogram (below) from same patient as in Figure 109. The lung scan shows complete absence of perfusion to the left lung and the arteriogram reveals intravascular filling defects in the pulmonary arteries to the RUL and RLL together with failure to opacify a number of segments of the left lung, presumably secondary to nonvisualized emboli.

Discussion

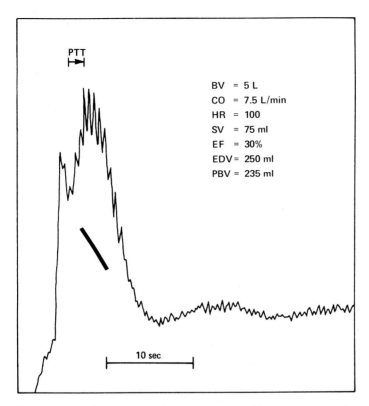

Fig. 109. Radiocardiogram from a patient with a recent massive pulmonary embolus showing greatly diminished pulmonary valley and markedly shortened PTT. This pattern is characteristic of markedly reduced pulmonary blood volume.

prominent LV portion of the record. The broken line in the figure is intended to indicate the washout pattern of the first passage of the bolus through the LV had the shunt not been present. Calculation of CO in such cases is not possible by exponential downslope analysis, but gamma-function analysis of the first portion of the LV peak may prove to be useful in such cases (21). The pattern of delayed washout of tracer from the LV can be distinguished from that seen in LV failure by noting that the first portion of the LV washout phase is sharp and by observing a cascading nature of the LV washout pattern. However, respiratory fluctuations superimposed on the pattern from a failing LV may superficially resemble the latter characteristic of a left-to-right shunt pattern.

DISCUSSION

While only substantially greater experience will establish the full clinical usefulness of this technique, two conclusions seem valid from information already available. First, the technique practiced at the bedside of the critically ill patient can, by qualitative means alone, assist in accurate differential diagnosis. Second, serial

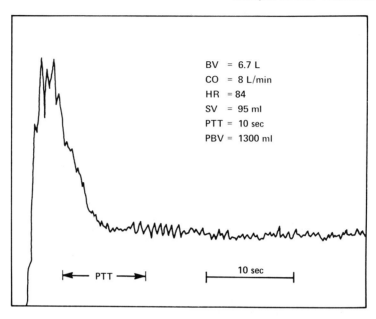

Fig. 110. Radiocardiogram of a patient with massive dilatation of the pulmonary arterial tree and moderate RV enlargement secondary to idiopathic pulmonary hypertension. Note the late appearance and low amplitude of the LV phase.

changes in the same patient in the same position should reveal even small changes, as a function of therapy or time, which give early and sensitive indication of central circulatory function. When the clinician seeks not a number but a guide to therapy, the single-proble study may occasionally or frequently provide all the necessary information. If indeed it does so, the benefits available to the patient in the hospital from radiocardiography are years closer at hand than if a gamma-camera and associated data storage, retrieval, and manipulation systems at the bedside are required. Quantitation is easily obtained and is in accord with angiographic measurements in the modest group of cases studied.

CRITIQUE

The need for a noninvasive, low-risk technique suitable for serial measurement of ventricular function is apparent in several clinical settings, ranging from patients not sick enough to require catheterization to patients too sick to be exposed to the risks thereof.

Outpatients evaluated for dyspnea, early congestive heart failure, the presence of an occult shunt, cardiomegaly of obscure cause, heart murmur of obscure origin, or other symptomatology suggestive of heart disease but not requiring hospitalization present a large group of patients whose diagnostic needs could best be met by a noninvasive method available with little advance notice and associated with no risk and negligible morbidity. The type of study defined in this chapter can generally be

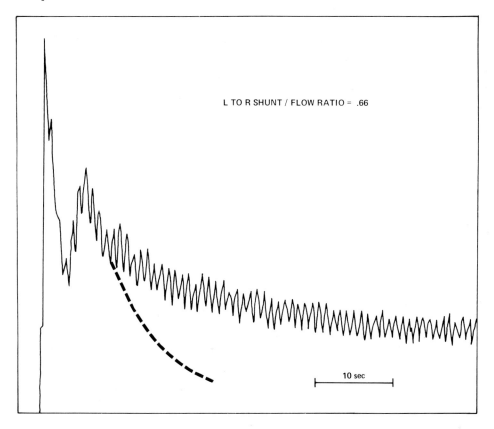

Fig. 111. Radiocardiogram from a patient with a left-to-right shunt through an atrial septal defect. The broken line in the figure is intended to indicate the washout pattern of the first passage of the bolus through the LV had the shunt not been present.

completed in less than 30 minutes of patient time, at an expense less than that incurred by 1 day of hospitalization, and at a patient radiation cost less than that received in any multiple-view radiographic procedure, such as an intravenous pyelogram or a barium enema. Furthermore, the diagnostic range of quantitative nuclear cardiography is broad enough to cover most of the cardiological spectrum, with the important exception of coronary artery disease and the less-frequent instances of idiopathic pulmonary hypertension. While other very important biophysical methods such as echocardiography, phonocardiography, vector cardiography, ballistocardiography, and perhaps roentgen kymography have specific and vital roles for one or another type of heart disease, none has so broad a potential to service an ambulatory clinic's diagnostic requirements. One might envisage a cardiac clinic in which selected patients would have a radiocardiographic study during the week prior to visit, or by special referral during a clinic visit, in much the way a patient has a follow-up x-ray or a blood glucose determination. Such studies would in essence be added to those now done, aiding in provision of quantitative data, but also probably adding to the total expense of patient care, since in most instances these patients would have no study if this were not done. Such development would be logical because it contributes to patient care, not because it reduces costs.

A second class of patients eligible for radiocardiography are those requiring cardiac catheterization for diagnosis of heart disease. Prior to the development of coronary-artery catheterization and coronary-artery surgery, most catheterizations were done for valvular and myocardial heart disease and for diagnosis of shunts. The great majority of catheterizations are now done for coronary-artery visualization, leaving in some instances a number of physiological studies at the bottom of waiting lists, or even unscheduled. The diagnostic spectrum of indications for catheterization indicates the potential "market" for radionuclide studies if the results were sufficiently diagnostic to replace the need for catheterizations done for other than coronary artery disease. Such a switch would require something akin to a religious conversion on the part of those who have developed the art of catheterization to its current high and precise state. However the ultimate place of each of these studies will be decided on cost/benefit ratios to the patient and the referring physician, whatever gradient of habit and preference may have to be overcome and however ingrained habit may delay the process.

A third category of patient may quickly benefit from nuclear cardiography. These are the intensive-care patients, either postoperative or in an acute and critical phase of response to a cardiovascular crisis such as myocardial infarction, papillary muscle or ventricular septum rupture, massive pulmonary embolism, or gram negative sepsis, who may benefit from drastic therapy, but who are too sick or too urgent for elective catheterization. While in some instances emergency catheterization is in the patient's best interest if it can be arranged, there are times when the simpler, less-invasive radionuclide study may provide the critical diagnostic data. Pierson (23) has described the results of studies done adjacent to a coronary care unit, in which serial measurements of cardiac output, ejection fraction, left ventricular volume, and pulmonary blood volume were done on 3 to 5 consecutive days in patients recovering from myocardial infarction. While the results of the study were trivial in this group of patients, none of whom developed anticipated complications, the feasibility of extending the measurement to the acute-care environment was demonstrated.

Will Quantitative Nuclear Cardiography Become Routine?

New procedures are accepted into cardiology ultimately according to rational process. Generally, do they vindicate their promise? Is the benefit/risk ratio favorable? Is the cost/benefit ratio competitive? And is it reliable under field conditions? Specifically these questions may be addressed by finding quantitative estimates for the sensitivity and specificity of each available diagnostic test for each cardiac lesion. In certain special diagnostic dilemmas, such as ruptured interventricular septum versus papillary muscle, myocardiopathy versus occult interatrial septal defect, and localization of a left-to-right shunt, the radionuclide test has special discriminatory competence. In other frequently encountered situations, such as late-stage mitral insufficiency versus myocardiopathy or cases in which coronary artery disease forms a part of the differential diagnosis, the greatly superior range of available measurements and of anatomic resolution provided by catheterization is required.

REFERENCES

1. M. L. Marcus, W. V. Schuette, and W. C. Whitehouse et al.: Automated Method for the Measurement of Ventricular Volume, *Circulation* **45**: 65-76, 1972.
2. D. C. Van Dyke, H. O. Anger, R. W. Sullivan, W. R. Vetter, Y. Yano, and H. G. Parker: Cardiac Evaluation from Radioisotope Dynamics, *J. Nucl. Med.* **13**: 585-592. 1972.
3. R. H. Jones, D. C. Sabiston, Jr., B. B. Bates, J. J. Morris, P. A. W. Anderson, and J. K. Goodrich: Quantitative Radio Nuclide Angiocardiography for Determination of Chamber to Chamber Cardiac Transit Times, *Am. J. Cardiology* **30**: 855-864, 1972.
4. J. A. Parker, R. H. Secker-Walker, R. Hill, E. J. Potchen, R. Siegel, and L. Resnick: The measurement of Left Ventricular Ejection Fraction Using a Scintillation Camera and a Small Digital Computer, *J. Nucl. Med.* **13**: 459, 1972. (Abstract)
5. J. P. Holt: Symposium on Measurement of Left Ventricular Volume, *Am. J. Cardiology* **18**: 208-225, 1966.
6. C. B. Mullins, D. T. Mason, W. L. Ashburn, and J. Ross, Jr.: Determination of Ventricular Volume by Radiosotope Angiography, *Am. J. Cardiology* **24**: 72-78, 1969.
7. H. W. Strauss, B. L. Zaret, P. J. Hurley, T. K. Natarajan, and P. H. Bertram: A Scintiphotographic Method for Measuring Left Ventricular Ejection Fraction in Man Without Cardiac Catheterization, *Am. J. Cardiology* **28**: 575-580, 1970.
8. R. W. Sullivan, D. A. Bergeron, W. R. Vetter, K. H. Hyatt, V. Haughton, and J. M. Vogel: Peripheral Venous Scintillation Angiocardiography in Determination of Left Ventricular Volume in Man, *Am. J. Cardiology* **28**: 563-567, 1971.
9. H. T. Dodge, H. Sandler, and R. E. Hay: An Angiocardiographic Method for Directly Determining Left Ventricular Stroke Volume in Man, *Circ. Res.* **11**: 739-745, 1962.
10. H. L. Blumgart, and S. Weiss: Studies on the Velocity of Blood Flow, *J. Clin. Invest.* **4**: 15-31, 1927.
11. J. Lammerant. P. Sprumont, and M. De Visscher: Enregistrement du Flot Sanguin Intra-Cardiac chez l'Homme par une Methode de Dilution d'Isotopes Radioactifs, *Arch. Intern. Physiol.* **64**: 65-71, 1956.
12. W. J. MacIntyre, W. H. Pritchard, T. W. Moir: The Determination of Cardiac Output by the Dilution Method Without Arterial Sampling, *Circulation* **18**: 1139-1146, 1958.
13. D. Parrish, D. T. Hayden, W. Garrett, and R. L. Huff: Analog Computer Analysis of Flow Characteristics and Volume of the Pulmonary Vascular Bed, *Circ. Res.* **7**: 746-752, 1952.
14. L. Donato, C. Giuntini, M. L. Lewis, J. Durand, D. F. Rochester, and A. Cournand: Quantitative Radiocardiography, *Circulation* **26**: 174-199, 1962.
15. R. Folse, and E. Braunwald: Determination of Fraction of Left Ventricular Volume Ejected per Beat and of Ventricular End Diastolic and Residual Volumes, *Circulation* **25**: 674-685, 1962.
16. R. J. Gorten: The Use of ^{125}Iodine for Precordial Counting, *J. Nucl. Med* **6**: 169-174, 1965; *ibid.*: Application of radionuclide procedures in clinical cardiology. AEC Symposium Series 27, E. Goswitz, G. Andrews, and A. Viamonte, Eds., National Technical Information Service, U.S. Department of Commerce, Springfield, Virginia 1971, pp. 594-613.
17. P. Steele, D. Van Dyke, R. S. Trow, H. O. Anger, and H. Davies: A Simple and Safe Bedside Method for Serial Measurement of Left Ventricular Ejection Fraction, Cardiac Output, and Pulmonary Blood Volume, *Brit. Heart J.* **36**: 122-131, 1974.
18. F. D. Sutton, P. P. Steele, D. Van Dyke, H. Davies, and J. H. Ellis: Patterns of Ventricular Function and Pulmonary Blood Volume in Acute and Stable Chronic Airways Obstruction, *Ann. Internal Med.*, Submitted for publication.
19. H. Parker, D. Van Dyke, P. Weber, H. Davies, P. Steele, and R. Sullivan: Evaluation of Central Circulatory Dynamics with the Radionuclide Angiocardiogram, in Diagnostic Nuclear Cardiology, H. Wagner, Ed., C. V. Mosby Co., St. Louis, Missouri, 1974.

20. R. H. Eich, W. R. Chaffee, and R. B. Chodos: Measurement of Central Blood Volume by External Monitoring, *Circ. Res.* **9:** 626–630, 1961.
21. L. Donato: Basic Concepts of Radiocardiography. *Seminar Nucl. Med.* **3:** 111–130, 1973.
22. R. L. Huff, D. Parrish, and W. Crockett: A Study of Circulatory Dynamics by Means of Crystal Radiation Detectors on the Anterior Thoracic Wall, *Circulation Res.* **5:** 395–400, 1957.
23. R. N. Pierson, Jr., D. H. Lin, J. Dagan, and F. S. Castellana: Serial Measurement of Left Ventricular Function by Quantitative Radiocardiography in a Coronary Care Unit, *J. Nucl. Med.* **14:** 438, 1973. (Abstract)

PRINCIPAL EDITOR: WILLIAM J. MACINTYRE

CONTRIBUTING EDITORS: PAUL J. CANNON AND WILLIAM W. ASHBURN

6. Measurements of Regional Myocardial Perfusion

Coronary heart disease constitutes a major public health problem in the world today. It has been estimated that 675,000 persons died and that 1,000,000 myocardial infarctions resulted from coronary atherosclerosis in the United States during 1968 (1). The development of surgical therapies for selected patients with angina pectoris, myocardial infarction, or heart failure secondary to coronary vascular disease (2, 3) has stimulated efforts to measure the regional blood flow distribution in the human myocardium.

During the course of evaluating patients suspected of having coronary artery disease, a number of diagnostic procedures are employed in order to characterize the state of myocardial perfusion. In the case of patients being considered for possible coronary bypass revascularization surgery, preoperative evaluation presently includes two necessarily invasive tests: (1) coronary arteriography: to locate and determine the extent of arterial narrowing and collateral flow and (2) left ventricular angiography: to evaluate the overall performance of the left ventricle as well as to search for regional abnormalities in ventricular wall motion which may be associated with ischemia, infarction, scarring, or aneurysm.

Although coronary arteriography provides excellent visualization of constrictions or occlusions of large- and medium-sized coronary vessels (4), the technique has four limitations: (1) radiographic visualization of the diseased artery does not provide information concerning the blood flow to the region of myocardium distal to the occluded segment; (2) arteriograms do not indicate whether there is occlusive disease of very small coronary vessels which are not visualized by current x-ray techniques; (3) if collateral blood vessels are not observed on the x-ray films, one cannot be certain whether or not there is a collateral blood supply to an ischemic cardiac region. Alternatively, if collateral vessels are visualized arteriographically, one cannot assess from the films alone whether or not the blood supplied by the collateral vessels is adequate to meet the tissue needs; (4) it is impossible to define critically or to investigate myocardial ischemia by radiography alone. Reports exist of patients with significant narrowings or occlusions of coronary vessels who did not have angina pectoris and of patients with typical angina pectoris and myocardial lactate production who had normal coronary arteriograms (5, 6). Current hypotheses suggest that angina pectoris and myocardial dysfunction occur when the coronary blood supply to a region of the myocardium is inadequate to meet its requirements for oxygen (7, 8).

If the "ideal" candidate for coronary bypass surgery is defined as one having a limited number of surgically approachable stenotic lesions with evidence of distal collateral flow and normal ventricular wall function, it would seem desirable to have a test that was capable of demonstrating the pattern of regional myocardial perfusion at the capillary or precapillary level, that is, beyond the limits of radiographic resolution. Such a test done pre- and postoperatively could also facilitate critical evaluation of the various procedures performed to revascularize ischemic heart muscle.

Furthermore, in patients with angina but without demonstrable arterial narrowing by arteriography, a measurement capable of evaluating quantitatively the overall distribution of myocardial perfusion might provide new insight into the pathological process.

Although the electrocardiogram is reasonably accurate in the diagnosis of 75 to 80% of infarctions, it provides only very partial evidence concerning the size, severity, and perhaps variation with time of the infarcted or ischemic area. Additionally, some areas of the heart are, in effect, electrically silent to standard electrocardiography. Finally, patients with bundle branch block, a frequent development in coronary artery patients, can rarely be assessed for location of ischemic or infarcted areas.

For these reasons considerable attention has been given to various radionuclide techniques that would provide an independent assessment of regional blood flow distribution in the normal, abnormal, or infarcted myocardium. Such measurements have been accomplished primarily by two general methods: (1) washout techniques that measure regional myocardial perfusion by determining the rate of clearance of an inert material such as radioxenon from the myocardium following intracoronary artery injection and (2) static imaging techniques which determine the relative deposition in the myocardium of radionuclides administered either in the form of particulate matter, which may be trapped in the capillaries following intracoronary injection, or as radionuclide in the form of diffusible materials, such as radiopotassium or potassium analogs, which may be administered either via a coronary-artery catheter or by peripheral venous injection.

While the latter technique has been used primarily for identification of ischemic or infarcted areas, it may theoretically be adapted for absolute flow measurements by quantifying either the rate of uptake or the fraction of dose administered.

Examples of each of these three methodologies will be shown in the following subsections of this chapter. A fourth methodology, static imaging with agents selectively concentrated in infarcted myocardium, familiarly known as "hot spot" imaging, is treated in Chapter X, which was added as this text was prepared for publication.

MEASUREMENT OF THE DISTRIBUTION OF MYOCARDIAL BLOOD FLOW USING ^{43}K AND DATA PROCESSING

A number of techniques measuring total myocardial blood flow have been devised using external counting devices to monitor the radioactivity in the myocardium following a single intravenous injection of ^{42}K, ^{86}Rb (9, 10), and ^{84}Rb (11, 12). These methods were originally proposed by Sapirstein (13) as *in vitro* procedures based on the indicator fractionation principle which postulates that the uptake of these tracers by the heart muscle equals the fraction of cardiac output perfusing the myocardium during the early stage of recirculation following a bolus of an intravenous injection of the tracer. This principle assumes that the extraction ratio of the myocardium during this stage does not differ significantly from that of the total body. Combining the myocardial uptake, total body uptake, and cardiac output, the myocardial blood flow can be calculated.

In the following investigation, relative values of myocardial blood flow in the various regions are of primary concern rather than total or absolute flow, although ultimately it would be desirable to express the regional distributions in terms of milliliters per minute. Scintillation-camera recording was chosen so as to view the entire myocardium simultaneously and thus minimize the effect of the washout phase that could be critical when one segment of the myocardium is measured at a different time than another segment. In addition, it was considered desirable to measure the distribution of the deposition in the myocardium during the short duration when the extraction ratio is predictable (14). In this situation, the scintillation camera would be more adaptable than the rectilinear scanner.

Since the gamma radiations of ^{42}K and ^{86}Rb are of too high energy for satisfactory scintillation-camera recording, ^{43}K with its primary gamma of 371 keV was chosen for these studies. In this study, ischemic areas were considered of greater concern than infarcted regions, which can be identified by more straightforward techniques. Thus ^{129}Cs was not utilized because of reduced extraction efficiency, slower myocardial clearance, and high blood background (15).

Method and Instrumentation

Patients were studied in the supine position with a scintillation camera placed over the precordium inclined to approximately 30° in the left anterior oblique position so that the camera face was perpendicular to the plane of the septum, hoping to separate the right heart from the left. The desired orientation is shown in Figure 112.

An additional elevation of 15° rotated around the transverse axis of the patient and inclined from the base to the apex of the heart was utilized to aid in separation of the atria from the ventricles. Since considerable penetration and scatter were noted with the conventional 3-in. 1000-hole collimator a $4\frac{1}{2}$-in. collimator (16) was used. Even with this collimator a long tail was observed on the line-spread function (17, 18) and compensation for the degradation of resolution by the interactive approximation method of Nagai and Iinuma (19) was employed.

With the patient in place under the camera, a dose of ^{43}KCl varying between

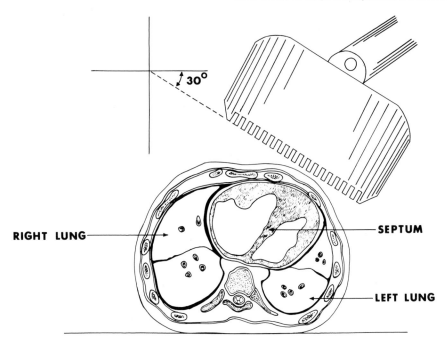

Fig. 112. Orientation of camera over the precordium. In addition to the rotation of 30° around the longitudinal axis of the body as shown, an additional 15° elevation of the camera face around the transverse axis was made, looking from base to apex of the heart, so as to help separate the atria from the ventricles.

0.7 and 1.2 mCi in a volume of 3 ml was rapidly injected into an antecubital vein followed by 10 ml of saline flush solution. The distribution of the material during the rapidly changing sequence was accumulated and transferred every 0.9 seconds as a digitized 40 × 40 matrix as previously described (Chapter 2).

At the end of 1 minute the frame accumulation time was increased to 60 seconds and recorded at this rate for the next 18 minutes. Following the recording of the last ^{43}K distribution, the spectrometer of the scintillation camera was changed to the ^{99m}Tc setting and a background count of the residual ^{43}K accepted by the ^{99m}Tc window was recorded. A rapid injection of 5 to 8 mCi of ^{99m}TcO$_4$ was then made by similar flushing techniques. The rapid phase was again recorded every 0.9 seconds until the monitor ratemeter showed the material clearing from the heart.

Results

The distribution of ^{43}K in the chest of a normal subject accumulated for 3 to 14 minutes following injection as shown in Figure 113*A*. The deposition in the left heart is greater than its surroundings and is easily visualized, even without data accentuation. The central decrease in the region of the left ventricle indicates that the desired orientation shown in Figure 112 has been achieved. This central decrease reflects the small amount of tissue directly over the chamber in contrast to the thick wall on the left supplied by the left coronary artery and the high deposi-

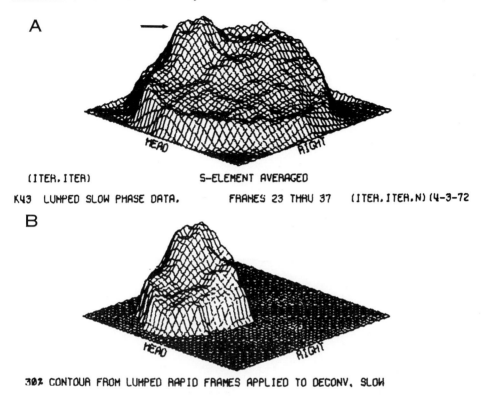

Fig. 113. (*A*) Deposition of ^{43}K in the chest of a normal patient 3 to 14 minutes after intravenous injection. Arrow designates deposition in the left ventricle with central decrease indicating region directly over the left-heart chamber. (*B*) Same record as *A* except that all elements outside the 30% contour line of the lumped frames of the ^{99m}TcO$_4^-$ passage have been set to zero. Note ratio of deposition of right to left heart is approximately 50%.

tion in the septal region supplied primarily by the left anterior descending branch. The right-heart area is less well delineated, however, and the entire outline of the heart is quite uncertain. In animal studies it had been feasible to outline the heart chambers by observing the passage of the bolus flow of ^{43}K as it passes through the heart (20). This was not satisfactory in human patients and thus the passage of ^{99m}TcO$^-_4$ as described in Chapter 2 was utilized to delineate the individual and combined chambers.

The outline of the heart obtained by the pertechnetate technique for the patient shown in Figure 113(*A*) can be superimposed on the matrix from which the three-dimensional display was derived. All elements falling outside this outline are considered to represent regions other than the heart, and thus these element values on the matrix are set to zero. The three-dimensional display of the resulting matrix is shown in Figure 113(*B*) which represents deposition of ^{43}K in those elements corresponding to a projection of the heart alone on the field of view of the camera.

The clinical usefulness of this technique is illustrated in Figure 114. The upper distribution of ^{43}K was recorded preoperatively on a patient with demonstrated occlusions of the right coronary artery and the left anterior descending artery. The

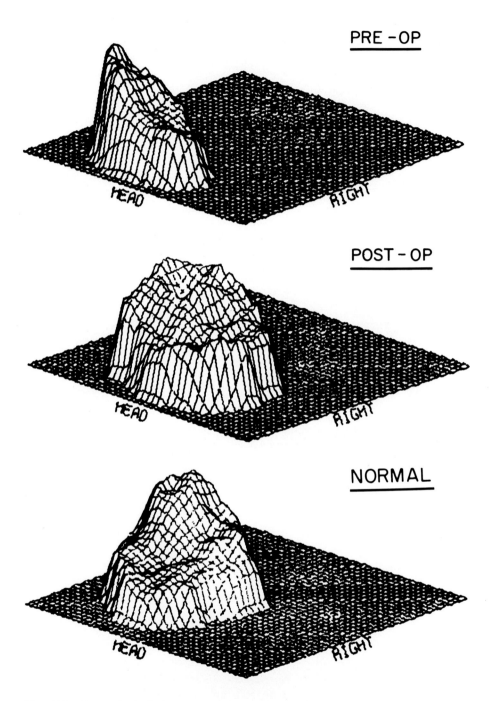

Fig. 114. Top and middle figures show the variation of ^{43}K in the myocardium before and after venous grafts successfully bypassing the obstructions of the right and left anterior descending coronary arteries. Bottom figure is the normal subject of Figure 113.

middle distribution shows the deposition 135 days later following venous grafts successfully bypassing the obstructions of the right and left anterior descending coronary arteries (21). This distribution shows an increased deposition of ^{43}K in the septal region that compares favorably with the normal distribution shown in the lower figure. The deposition in the right heart is also markedly increased over the preoperative distribution.

While it would be expected that the resolution of detail derived from direct injection of the radioactive material into the coronary arteries would be superior to that derived from a peripheral intravenous injection, it is of interest to examine how closely the two techniques do agree. Figure 115A shows the distribution of ^{43}K recorded following injection into the right and left coronary arteries of a patient with a known complete occlusion of the left anterior descending artery at its origin. In this type of injection, little material is deposited elsewhere, and Figure 115(B) shows only minor change after the elements in the extracardiac regions are eliminated. Both demonstrate the large central defect in the septal region due to the occlusion.

Figures 115(C) and 115(D) show the same patient studied following injection of ^{43}K into a peripheral vein. The right-heart deposition is appreciably higher since in the previous study Figs. 115(A and B) separate injections into the right and left coronary artery were made and the amount delivered to the right side was low. It is realized that the intracoronary injections will seldom yield the same ratio of deposition in the right and left heart as would be obtained by the intravenous injection. Normally, twice the amount of radioactivity is injected into the left coronary artery as is injected into the right. The fraction of the dose actually delivered to the desired region, however, can be quite variable. The same central septal defect is noted, however, in Figures 115(A) and 115(C) even though the heart outline in Figure 115(B) is much more difficult to ascertain than the outline in Figure 115(D).

Preliminary results in 19 patients in whom the information derived from the angiograms has been compared with the information obtained from the deposition of ^{43}K injected directly into the coronary artery have indicated a reasonable correlation in addition to supplementary information being supplied by the ^{43}K perfusion (22).

Future Clinical Uses

There is still a question, however, as to the limitations that may prevent similar information from being derived from the atraumatic intravenous injection of ^{43}K. When the passage of ^{99m}Tc through the large chambers is reasonably fast, the chambers can be well outlined and, as was shown in Figure 115, the results are comparable. In cases of heart failure or any prolonged slow passage, the identification of the left heart has been difficult and it is possible that other techniques may be required for these patients.

The extension of these measurements to express absolute rather than relative myocardial flow will necessitate further refinement. Subtraction of the contribution of the radioactivity in the blood by the method of Donato et al (9, 10)

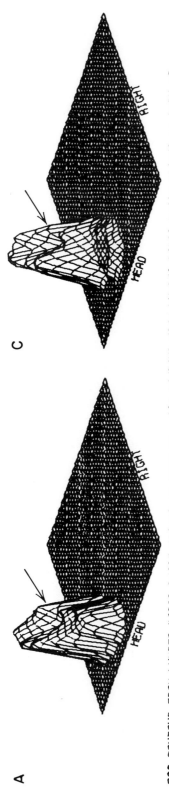
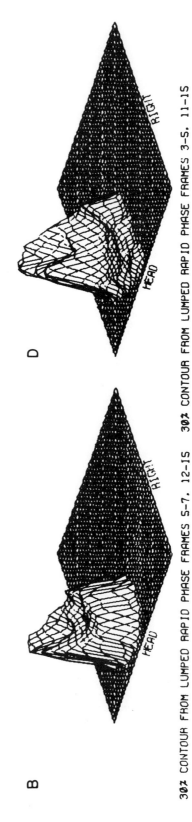

Fig. 115. (A) Three-dimensional representation of deposition of ^{43}K in the myocardium following injection into the coronary arteries. Arrow designates septal region with decreased deposition due to occlusion of the left anterior descending artery. In the matrix from which this figure was drawn, all elements outside of the heart area have been placed at zero. (B) Full representation of precordial scan following injection into the coronary arteries. (C) Similar representation as A except that the ^{43}K has been injected into a peripheral vein. Arrow shows decreased deposition in septal region. (D) Full representation of precordial scan following intravenous injection.

can, of course, be achieved by substituting ^{99m}Tc-tagged albumin rather than using the pertechnetate ion. We have usually integrated the accumulation of ^{43}K in the myocardium between 3 and 10 minutes after injection. It will be necessary to ascertain the relationship of the integrated extraction ratio of the myocardium to the whole body during this period. Lastly, the effect of the uptake of ^{43}K by the back muscles, insignificant in the animal studies, will require correction in patient determinations in order to express the results in absolute flow.

In terms of *relative* myocardial blood distribution, experience with 29 patients injected intravenously with ^{43}K has indicated the method to be relatively simple and productive of useful clinical information. Its use to evaluate such procedures as coronary vein bypass grafts (Fig. 114) is of great immediate interest, but also of concern in the possibility of early evaluation of myocardial infarction and following the progress of revascularization. Since the procedure is relatively nontraumatic, its usefulness as a measurement of changes of regional myocardial flow with drugs, exercise, diet, age, or other factors would merit serious consideration.

MYOCARDIAL PERFUSION IMAGING WITH LABELED PARTICLES INJECTED INTO THE CORONARY CIRCULATION

Considerable successful experience in identifying abnormal regional pulmonary perfusion patterns with labeled macroparticles has led to the concept that a similar technique might be applied to patients with abnormalities of myocardial perfusion. Accordingly, several groups of investigators (23–25) have studied the feasibility of injecting radioactive particles directly into the coronary circulation at the time of coronary arteriography. This was based on the original work in dogs by Quinn et al. (26).

Toxicity studies in dogs (27–30) have shown that when the number of particles is kept below 500,000 and the size is carefully controlled (10–60 microns) no significant alterations in cardiac performance resulted. In a survey of those centers employing this technique, over 1200 patients, many of them with abnormal coronary circulation, have received such intracoronary injections of labeled particles with no reported mortality or significant morbidity.

Method of Injection

Ashburn (24) has reported a technique in which two different radionuclides are employed in order to study selectively the right and left coronary arterial distributions. Approximately 300,000 sterile denatured albumin macroaggregates labeled with ^{131}I and diluted in 5 ml of sterile heparinized saline are slowly injected by hand into the right coronary artery through a Sones catheter. The position of the catheter tip both before and after injection of the particles is confirmed by injecting a small quantity of contrast material through the catheter. Following this, the left coronary artery is similarly injected with approximately 150,000 particles labeled with ^{99m}Tc. Since the number of particles is held constant, the amount of radionuclide injected varies depending upon the available specific activity of the labeled product. Thus, 50 to 85 microcuries of ^{131}I and 500 to 1500 μCi of ^{99m}Tc are injected per study. Each batch of particles is carefully inspected under the microscope for uniformity in size and the approximate number of particles per cubic millimeter is determined. Any product containing particles larger than 80 μ is discarded. Generally, particle size ranges between 10 and 60 μ in diameter. Immediately prior to administration, the microspheres are reultrasonicated and inspected under the microscope for uniformity, and a rough count is made in a hemocytometer. The desired quantity is withdrawn into a 1-ml tuberculin syringe and diluted to 1 ml with sterile heparinized saline. Following preliminary review of the coronary arteriograms, the desired quantity of particles is transferred to each of two larger syringes (containing approximately 5 ml of heparinized saline) according to the desired distribution (e.g., 60% or 0.6 ml for the left coronary artery injection and 40% or 0.4 ml for the right), remembering that each 0.1 ml in the tuberculin syringe contains approximately 10,000 labeled particles.

The dual-label technique has the advantage of permitting separate appraisal of the precapillary distribution of myocardial blood flow from the right and left coronary arteries. The principle disadvantage lies in the fact that the higher energy of the gamma photons of ^{131}I requires compromises in collimator selection, particularly with the scintillation camera, which results in less than optimal image resolution and sensitivity. Furthermore, the low count rate associated with the small dose of ^{131}I (50 μCi average) often necessitates examination times in excess of 10 minutes per view to obtain 40,000 counts. This is in contrast to the count rate from ^{99m}Tc which permits 200,000 counts and a superior image to be recorded in 5 minutes.

A more recent modification of the original technique has been advocated by Weller et al. (29) in which ^{99m}Tc-labeled human albumin microspheres ranging in size between 15 and 30 μ are injected into both the right and left coronary arteries. Current technique consists of injecting a total of no more than 100,000 microspheres into the coronary circulation, the dose being divided between the left and right coronary arteries, depending on the radiographically determined coronary anatomy; for example, in a dominant left coronary circulation, approximately 60,000 microspheres are injected into the left coronary artery and 40,000 microspheres into the right coronary artery.

In evaluating the perfusion pattern following a saphenous vein coronary arterial bypass graft, the particles may be injected directly into the graft. In this situation, a convenient method is to inject ^{131}I-labeled particles into the graft and ^{99m}Tc microspheres into both the right and left coronary arteries.

Following labeled-particle administration, the cardiac catheterization procedure is usually terminated. Since the clearance rate of the labeled particles from the coronary vascular tree has a half time in the range of 3 hours, myocardial imaging may be delayed if necessary up to several hours, although the imaging procedure can normally be begun within 30 minutes of the injection. This allows ample time to close the arteriotomy site and monitor the patient's vital signs before transfer to the nuclear medicine facility. As a rule, particles have not been injected in those patients experiencing more than mild or transient angina or ECG alterations associated with the injection of radiographic contrast material into the coronary circulation, although these patients may have the most to gain from the procedure.

Imaging

Imaging of the myocardial distribution may be accomplished with either a rectilinear scanner or scintillation camera. It is essential that multiple views be obtained; these should include the anterior and left lateral as a minimum, with both oblique and posterior projections added if possible. Pulse-height discrimination for the gamma-ray energies of ^{131}I and ^{99m}Tc permits separation of the perfusion patterns of the right and left coronary arteries when particles labeled with these radionuclides are used. Imaging devices not having dual pulse-height capabilities require two exposures per view, for example, 40,000 or more counts (^{131}I) and 200,000 or more counts (^{99m}Tc).

When ^{99m}Tc microspheres are injected into both coronary arteries, higher resolution and shorter exposure times are possible. Imaging is ideally performed with a scintillation camera using a low-energy converging or high-resolution collimator. A minimum of 200,000 counts per view are desirable for optimal images. Since the heart is beating during the period of exposure, electronically gating the data acquisition with the ECG to include only the 100 msec corresponding to end diastole (QRS complex of the ECG) does noticeably improve resolution (see Figure 115), but necessarily increases exposure time by a factor of approximately 10, depending on the heart rate and width of the gate.

Standard analog imaging on Polaroid or any one of the various types of negative film, such as 70 mm or x-ray, is acceptable. However, if ECG gating in a particular view is desired, simultaneous recording of both the ECG and the scintillation-camera positional data in a real-time digital recording system is recommended. Direct recording with a small dedicated computer system may be useful, allowing correction for inhomogeneities in scintillation-camera response, data-point averaging, and enhancement of edges. While it is entirely within the capability of the computer to combine separate exposures from comparable views (e.g., the anterior ^{131}I and ^{99m}Tc images), such addition, whether or not color addition is employed (31), requires correction for the relative efficiencies of detection for each radionuclide, the total counts accumulated per exposure, and some knowledge of the relative number of particles injected into each vessel as well as the relative coronary blood flow of the right and left coronary arteries. This, at best, can only be roughly approximated.

Interpretation

Two dimensional perfusion images are the result of the distribution of radioactivity in a three-dimensional structure. Overlaps of the ventricular septum upon the ventricular wall, the anterior wall upon the posterior wall, and normally perfused regions superimposed upon underperfused areas complicate interpretation, and careful analysis and correlation of all views is absolutely necessary. The ventricular cavities which, of course, do not contain radioactivity are often well outlined by the activity in the surrounding ventricular walls in a "doughnut" effect. The relatively greater thickness of the left ventricular wall as compared with the right and the

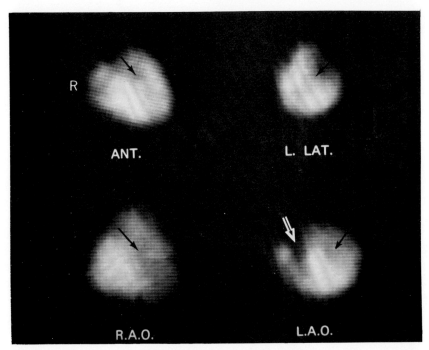

Fig. 116. Myocardial perfusion images in a patient with no evidence of coronary artery disease. ^{99m}Tc HAM was injected into both coronary arteries (40% into right coronary artery and 60% into left—see text). The relative void in radioactivity corresponding to the left ventricular cavity can be appreciated in all four projections (black arrows). The right ventricular cavity is best seen in the LAO view (black and white arrow). Note the prominent accumulation in the ventricular septum superimposed upon the left ventricular wall. This and all subsequent images (Figs. 117–119) were generated in a 64 × 64 data point matrix from previously recorded data in a digital computer system. ANT, anterior; L. LAT, left lateral; RAO, right anterior oblique; LAO, left anterior oblique.

increased myocardial blood flow to the left ventricular wall are easily distinguished in high-resolution images (Fig. 116). In most cases, the accumulation of labeled particles in the atrial walls can be seen only if the ventricular images are purposely overexposed.

Underperfused regions are identified by focal areas of reduced or uneven accumulation of the labeled particles (Fig. 117 and 118). The relative size and location of these defects can be best analyzed by observing the abnormalities in multiple projections. A small or partially underperfused area may be easily overlooked by superimposition of the defect upon the ventricular cavity in one projection, but identified in other views. High-resolution ECG gated images (e.g., end systolic and end diastolic) in the appropriate projections, besides improving detail, may also provide evidence of the relative motion of the myocardial tissues surrounding the defect during the cardiac cycle (Fig. 119).

Results

Clinical experience with myocardial aggregate perfusion imaging suggests that when care is taken to assure that the labeled particles are of the correct size and number,

Myocardial Perfusion Imaging with Labeled Particles Injected into the Coronary Circulation

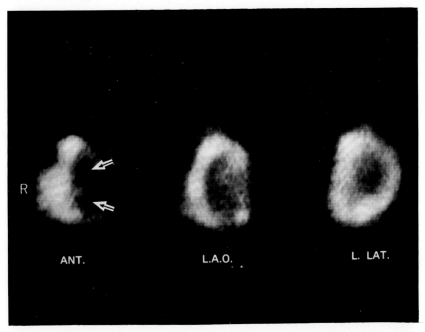

Fig. 117. Patient with obstructions in the distal right coronary and proximal anterior descending branch of the left coronary artery. A well-defined region (arrows) of reduced perfusion in the anterolateral aspect of the left ventricular wall (associated with dyskinesis of the left ventricular wall on contrast angiography) is easily identified in all four views, but its extent and location are best determined by correlation of all views. ^{131}I-albumin macroaggregates were injected into the right coronary artery and ^{99m}Tc HAM was injected into the left coronary artery.

no untoward effects or observable alterations in left ventricular function have been encountered.

Myocardial perfusion imaging has been comparable with the results of left ventricular angiography in virtually all cases. Rarely, if ever, has a region of reduced uptake in multiple views not been associated with an area of abnormal wall motion in the same location, whether this region was hypo- or akinetic or showed paradoxical movement. In general, the area of involvement, that is, relative hypoperfusion, appears on the radionuclide image to be somewhat smaller than that suggested by the angiographic extent of abnormal wall motion. Several discrete and/or confluent areas of relative underperfusion may be defined by the radionuclide technique, while routine analysis (i.e., without the aid of video tracking or similar techniques) of the cine left ventricular angiogram often fails to suggest the patchy involvement.

Correlation of the results of myocardial perfusion imaging with coronary arteriography suggests that obstructions of major branches exceeding 70% are usually associated with recognizable regional hypoperfusion on the appropriate radionuclide images. This is more marked when there is a relative lack of collateral perfusion, as when multiple vessels are involved and in those cases in which extensive scarring of the myocardium has occurred. Some instances of regional hypoperfusion have been noted by radionuclide imaging when the original interpretation of the cine coronary arteriogram has been normal. Usually, upon reexamination of the

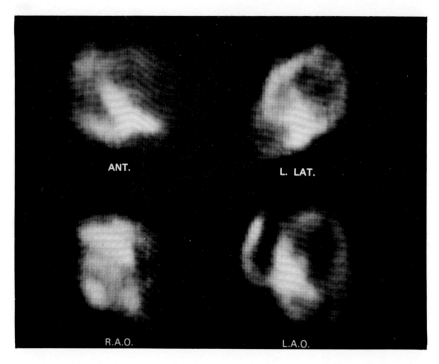

Fig. 118. Obstructions in all major branches associated with multiple regional perfusion deficits. ^{99m}Tc HAM was injected into both coronary arteries (40% right and 60% left coronary).

contrast study, obstructions of smaller vessels, often at their origin, or a paucity of vessels supplying a region of previously infarcted myocardium with scarring are discovered.

It would be premature to define the ultimate place of myocardial perfusion imaging in the selection of suitable candidates for myocardial revascularization or coronary bypass procedures. However, it is clear that interpretation of the images depends on careful correlation with coronary arteriography and left ventriculography. Judkins has suggested that symptomatic patients with surgically approachable coronary arterial lesions having normal direct or collateral flow, as evidenced by a relatively normal distribution of myocardial blood flow on the perfusion images, are better risks for revascularization surgery than patients in whom regional obliteration or reduction in small-vessel perfusion is shown by the radionuclide study. Similarly, a normal coronary arteriogram associated with normal left ventricular wall motion and a normal radionuclide perfusion pattern is reassuring with respect to the state of both the large and small coronary vessels. The incidence of normal arteriography with abnormal perfusion images is not yet known. Nor is the probability that bypass would be of value in such a patient. Considerably greater experience with myocardial perfusion imaging in conjunction with contrast arteriography and other indirect indicators of myocardial function will be needed before determining the ultimate significance and value of the radionuclide technique.

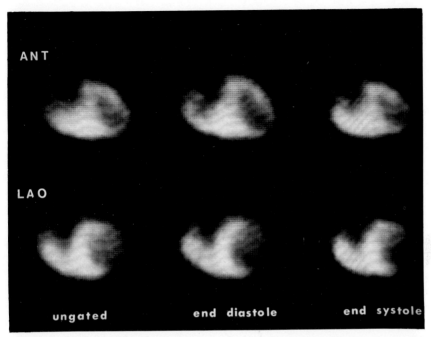

Fig. 119. Improvement in resolving the edges of a perfusion defect by gating the images with the ECG (see text) can be seen in comparison with the ungated images. All images consist of 100,000 counts. The patient had obstructions in the distal right coronary artery and proximal left anterior descending and circumflex branches of the left coronary artery.

Unresolved Problems

One of the rather obvious criticisms of the technique of injecting labeled particles directly into the coronary circulation is that complete mixing of the particles with coronary arterial blood is unlikely. The results of streaming of the contrast material due to the jet effect associated with injection through a single-hole arterial catheter are well known. Indeed, we have observed that a rapid injection of labeled particles into the left coronary through a Sones catheter resulted in what would appear to have been a virtual absence of perfusion of the distribution of the left anterior descending branch in spite of a normal coronary arteriogram. This has occurred only once and was probably related to the rapid rate of particle injection which was easily recognized by a similar streaming effect of the contrast material flowing primarily to the circumflex branch during the test injection. This problem would appear to be minimized by careful placement of the catheter and by a slower rate of particle administration. In this regard, a comparison was made in four patients between the myocardial images obtained following the slow administration of ^{99m}Tc-labeled particles into the left coronary artery and the images following a similar injection of ^{43}K after the reinsertion of the catheter into the left coronary orifice. Excellent correlation in the regional distribution of the two agents was found in all the cases. This would suggest that (*a*) although the mechanism of extraction of these agents is different, the images are probably comparable (within the limits of attainable resolution) and (*b*) the method of direct intracoronary injection employed is reproducible with respect to image consistency and probably is not greatly affected by catheter position and/or streaming.

Of perhaps greater concern is the inexactness with which the total dose is divided between the right and left coronary arteries during the separate injections. Studies performed in our laboratory in which a predetermined quantity of ^{99m}Tc-labeled particles was selectively injected into both coronary arteries according to our angiographic impressions of the relative perfusion via each major coronary artery were compared with the images produced in the same patients following the intravenous administration of ^{129}Cs. Good agreement was found in most cases, suggesting that our decision as to the relative quantity of labeled particles to inject into each coronary artery was reasonably correct. The most dramatic exception to this was shown (Fig. 120) in a patient in whom the degree of occlusion in the upper third of the right coronary artery was not fully appreciated. This resulted in a disproportionately large quantity of labeled particles being injected into the right coronary artery with an exaggeration of the uptake in the right atrial appendage and other myocardial tissue perfused by the proximal branches of this vessel. Even so, a close **examination** of the quality of the ^{129}Cs image in this example demonstrates the difficulties encountered in myocardial imaging with this agent; specifically, the high tissue background levels, scatter, and the resulting poorer resolution.

THE MEASUREMENT OF REGIONAL MYOCARDIAL PERFUSION WITH ^{133}XENON AND A SCINTILLATION CAMERA

The method of measurement of regional myocardial perfusion consists of injection of ^{133}Xe selectively via a catheter into a coronary artery and recording isotope washout curves from multiple discrete areas of the myocardium with a multiple-crystal scintillation camera (32–36). Rate constants of isotope washout from heart muscle are calculated using a monoexponential model, and the myocardial capillary blood flow rates in multiple regions of the heart are calculated by the Kety-Schmidt formula. The pattern of regional myocardial perfusion rates so obtained is superimposed upon a tracing of the patient's coronary arteriogram obtained in the same study. Using this method the pattern of myocardial perfusion rates observed in normotensive patients with normal coronary arteriograms has been described (35), and significant alterations in myocardial perfusion have been found in patients with arteriographically apparent coronary artery disease (36).

During the past two decades many investigators have measured the clearance rate of inert gas from the myocardium in order to estimate the average myocardial blood flow per gram tissue in a large area of the heart (a ventricle or more) in animals (37–39) and man (38, 40, 41). After the myocardium had been loaded with inert gas through inhalation or intracoronary injection, single inert-gas washout curves were obtained either by taking multiple samples of coronary sinus blood or by external precordial counting with a single scintillation detector. However, when these measurement techniques were applied to the study of patients with coronary artery disease, the data have not been clinically useful. Blood flow values

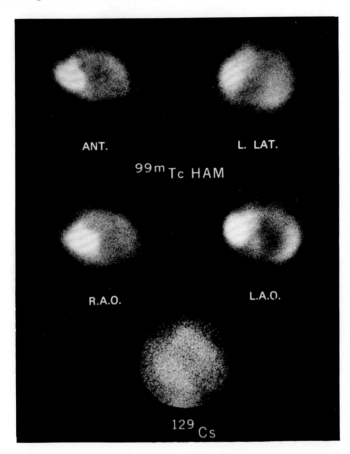

Fig. 120. Patient with complete obstruction in the proximal right coronary artery as well as an obstruction in the anterior descending branch of the left coronary associated with apical dyskinesis. ^{99m}Tc HAM was injected into both coronary arteries with 40% injected into the right coronary and 60% into the left. Thus, 40% of the total dose became trapped in the branches of the right coronary proximal to the obstruction resulting in a fallaciously intense uptake in the upper portions of the right heart. This is better appreciated by comparing the anterior particles perfusion image with an anterior image obtained following the intravenous injection of ^{129}Cs.

comparable to normal were found in patients with significant abnormalities on the coronary arteriograms (38, 40). This probably occurred because small areas of reduced myocardial blood flow were obscured by much larger regions with normal perfusion when only a single myocardial washout curve was measured (38). In studies with H_2 (41) the single inert gas washout curve obtained from patients with coronary artery disease was found to deviate significantly from a single exponential, but in a fashion which could not be accurately resolved mathematically to give regional myocardial blood flow data.

Indicator

In the present studies the tracer used to estimate regional myocardial perfusion was ^{133}Xe (35), a radioactive isotope of a gaseous element which is chemically inert, physiologically inactive, and highly diffusible such that its passage across capillary walls into the tissue supplied by an artery is rapid and is not limited by diffusion through pores. The washout of the tracer from heart, brain, kidney, and other tissues has been shown to be a function of the tissue capillary blood flow (37–39, 51–53). The solubility coefficient between air and blood is 9:1. Because more than 95% of the isotope is eliminated to air in one passage through the lungs (54), recirculation of tracer is not a major problem in blood-flow measurements and the radiation exposure of the patient is reduced. An intravenous or intra-arterial dose of 40 mCi of ^{133}Xe dissolved in saline gives a gonadal dose of 10 mrads, a level considered acceptable for routine diagnostic isotope work (55).

Inert-Gas Measurements of Nutrient Blood Flow

The inert-gas techniques to measure nutrient capillary blood flow per unit mass of tissue which were developed by Kety and Schmidt have been reviewed extensively (56, 57). The assumptions inherent in this approach are listed in Table 4. With regard to the myocardial circulation, it is known that coronary flow is phasic; however, assumption 1 stipulates only that averaged arterial inflow and venous outflow be constant throughout the period of measurement. The second assumption is that the inert gas equilibrates instantaneously with the perfused tissue; in this regard studies using flow meters have provided data indicating that the myocardial washout of inert gases is not diffusion limited, but a function of blood flow (39). Also, scintillation-camera studies showing very rapid spread of labeled inert gas through tissue beds supplied by the artery into which the tracer was injected (32) provide additional support for the assumption of rapid equilibration between blood and perfused cells. Sizeable arteriovenous shunting in the perfused organ could prevent equilibration of labeled tracer in the tissue and invalidate inert-gas measurements of tissue capillary blood flow; microsphere studies, however, have indicated that arteriovenous shunts through the myocardium do not exist (42, 58). The inert-gas methods also assume homogeneous perfusion in one or more compartments of flow in the tissue bed under study. The presence of a countercurrent exchange vascular

Table 4. Assumptions of Kety-Schmidt Method for Measuring Regional Blood Flow

1. During the measurement, arterial inflow and venous outflow are equal.
2. Instantaneous equilibrium of indicator between tissue and blood according to its partition coefficient is attained.
3. Homogeneous perfusion exists in tissue bed under study (absence of a countercurrent exchange vascular system).
4. The indicator is not metabolized; it is removed only via venous blood and does not recirculate.

system (such as occurs in the vasa rectae of the renal medulla) may invalidate calculations of blood flow from inert-gas washout curves since trapping of gas by the system causes its rate of washout to be slower than that predicted from the capillary blood flow rate. Recent studies by Yipintsoi and Bassingthwaighte, using perfused dog hearts, showed that at low rates tissue tracer uptake curves had a more steeply rising up-slope with tritiated water than with ^{125}I-antipyrine when both tracers were injected simultaneously into the coronary artery (59, 60). This observation, plus the finding that in similar studies with ^{133}Xe (39) fractional escape rates from blood to tissue were higher than expected from simultaneous ^{125}I-antipyrine curves, led to the postulate that within deeper regions of heart muscle some diffusional shunting across arteriolar or capillary loops may occur (59). The magnitude of this effect, as indicated by the transit times (60), is very small and as such would be unlikely to affect a washout measurement. This suggestion is supported by observations that inert-gas measurements of flow per gram times heart weight have correlated well with flowmeter myocardial flow rates (37–39, 61). With regard to assumption 4, there is no evidence that inert gases are metabolized; removal by cardiac lymph, if present, must be negligible in relation to removal by blood flow. Studies have indicated that >95% of ^{133}Xe or ^{85}Kr is exhaled after one passage through the lungs (54); however, studies after injection of these labeled gases into a variety of organs indicate that some small amount of recirculation does occur and can modify later portions of inert-gas washout curves (51, 52).

Theoretical Considerations

The Kety application of the Fick principle may be expressed as:

$$\frac{dQ_i}{dt} = F(C_a - C_v) \tag{1}$$

where Q_i is the amount of radioactive inert gas in the myocardium, C_a and C_v are the concentrations of gas in the arterial and venous blood respectively, t is time, and F is blood flow. After injection into an artery of a bolus of a radioactive gas which does not recirculate, the arterial concentration, C_a, becomes 0. The venous concentration, C_v, can be expressed by $C_v = Q_i/V\lambda$ where V equals the volume of distribution of the gas and λ the partition coefficient describing the relative solubility of the gas in blood and myocardium at equilibrium.

Thus, Equation 1 can be rewritten:

$$\frac{dQ_i}{dt} = \left(\frac{Q_i}{V}\right) \tag{2}$$

This can be rearranged as

$$\frac{dQ_i}{Q_i} = \left(\frac{-F}{V\lambda}\right) dt \tag{3}$$

The mathematical expression of exponential washout from a homogeneous compartment is

$$A_t = A_0 e^{-kt} \tag{4}$$

where A_0 is the initial amount, A_t the amount at time (t), e the natural logarithm, and k the rate constant of washout.
Thus,

$$Q_{it} = Q_{i0}\, e^{-(F/V\lambda)t} \qquad (5)$$

where Q_{it} and Q_{i0} equal the amounts of radioactive gas at times t and 0, respectively. However,

$$\frac{F}{V\lambda} = k \qquad (6)$$

the rate constant of tracer washout which can be determined experimentally. Therefore,

$$\frac{F}{V} = k \times \lambda \qquad (7)$$

However, V can be expressed as W/ρ where w is the weight of the tissue and ρ is the specific gravity (1.05).

Since the weight of the tissue is not known, by convention flow is expressed in terms of 100 g of tissue; thus 100 is substituted for W and flow is expressed as milliliters per 100 g myocardium per minute.

$$F = \frac{k \times}{\rho} \times 100 \qquad (8)$$

Instrumentation

The development of gamma-ray scintillation cameras (62, 63) made possible the present attempt to apply the inert gas measurement technique to the study of regional tissue blood flow measurements within the heart. The following studies were performed using the multiple-crystal scintillation camera* and a multichannel collimator (33–36). Static studies with the $1\frac{1}{2}$-in. multichannel collimator used in the present studies indicated that the radius of myocardium viewed by each crystal was 6 mm at 4 cm distance (corresponding to the anterior cardiac wall) and 12 mm at 8 to 9 cm (corresponding to the posterior cardiac wall).

Figure 121 shows that beyond a certain distance from the collimator the field of view of a crystal will overlap that of an adjacent crystal. The degree to which there is overlap of field of view is determined by the characteristics of the multichannel collimator (64) and, in general, diminishes with increasing septal thickness. If there is significant overlap of field of view of adjacent crystals (or areas within a large single crystal), radiation arising in one region of the myocardium will be recorded by a crystal overlying another region. This is of major importance in dynamic studies of isotope removal from adjoining regions; calculations (35) have indicated that serious errors can be introduced into measurements of regional myocardial blood flow if there is a significant percentage of counts recorded by a detector overlying one region of the heart (e.g., with rapid blood flow) which arise from

---

* Baird Atomic Autofluoroscope Model 5600.

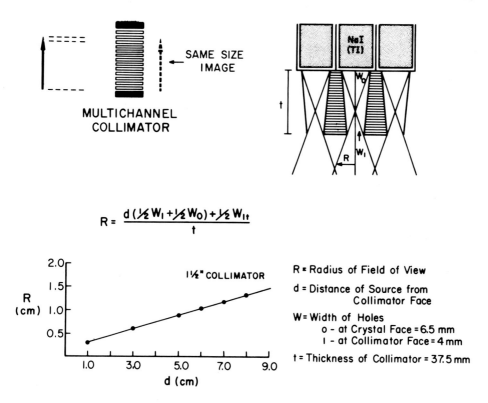

Fig. 121. Effective resolution of each scintillation detector of the autofluoroscope is determined by the geometry of the channels in the collimator and the distance of the collimator face from the ^{133}Xe source.

an adjoining region (e.g., with a slower blood flow rate).* In addition, overlap, whether occurring in a multicrystal- or single-crystal camera, results in smoothing or averaging of resultant data such that a flow map resulting from the data blurs sharp contiguous discrepancies and hides small anomalous areas that differ sharply from surrounding regions.

Static studies with the collimator used for these studies indicated that the magnitude of overlap of the field of view of adjacent crystals was negligible ($<3\%$) at distances up to 5 cm from the collimator face, but rose to 16.4% at a distance of 8 cm.

Technique

Selective coronary arteriography is performed according to the technique of Sones or by a modified Judkins approach (4, 65). Following the selective arteriography

---

* This problem also exists if attempts are made to record regional isotope clearances with a single-crystal camera and a multichannel collimator. If the collimator used does not preclude such overlap, regional clearance data obtained by subdividing data recorded by one large scintillation crystal on a computer will contain this error.

of the artery under study, assuring appropriate catheter placement, the patient is placed in the left anterior oblique position. A series of radiopaque–radioactive markers are positioned on the chest wall to indicate the coronary ostia and heart borders. A small amount of contrast is injected into the artery and another cineangiogram taken to indicate the positions of the markers relative to the coronary artery under study (at least three markers oriented in two axes are filmed at a known magnification along with the coronary artery). After 5 to 10 minutes, to allow possible vasoactive effects of the contrast medium to be dissipated, the angiographic camera is replaced by the multiple-crystal scintillation camera. The collimator face is positioned over the patient's precordium in a position identical to that of the image intensifier; the positions of the radioactive markers are recorded onto the magnetic tape, and the markers removed. (Fig. 122).

Twenty to 25 mCi of ^{133}Xe dissolved in 1 to 3 ml of sterile pyrogen-free isotonic saline is injected rapidly (2 to 3 seconds) through the catheter into the main right or left coronary artery, followed immediately by a flush of 2 to 4 ml of saline. Gamma radiation emitted by the isotope as it diffuses into myocardial cells and is subsequently washed out as a function of myocardial capillary blood flow is monitored by each of the scintillation crystals overlying the heart; the data from all 294 crystals are recorded as counts per second onto magnetic tape at 1/second framing rates for 4 to 7 minutes. ^{133}Xenon exhaled by the patient may be trapped or vented to the outside air by a vacuum exhaust system held close to the patient's face. Both coronary arteries may be studied in this manner.

Figures 123a and 123b show scintiphotographs obtained by replaying data from normal right and left coronary artery studies onto an oscilloscope. The dynamic radioisotope images generated during arrival and washout from the areas of myocardium perfused by each coronary artery correspond in size and shape to the area of myocardium which receives its capillary blood supply from the injected artery.

Data Processing

The data from each study must be processed by a digital computer. Corrections are programmed for: (a) the slightly differing efficiencies of the 294 crystals (assessed by counting a uniform pool of ^{57}Co), (b) the dead-time required by the instrument to electronically process observed radioactive events (experimentally determined to be 15 μsec for the instrument used in these studies), and (c) the time required to transfer the matrix onto magnetic tape (45 msec). Each frame of data or calculations is printed in 21 rows of 14 numbers with each number in the printout corresponding to the matrix location of the crystal from which the information was obtained.

Crystals which overlie the myocardial tissue supplied by the artery into which ^{133}Xe was injected are identified by the computer which examines the isotope washout curve recorded by each crystal and prints out the peak observed count rate and also the number of seconds after the start of the study that the peak occurred (Fig. 124). The peak count rates in crystals overlying the myocardium are higher and occur sooner than in crystals recording excretion of ^{133}Xe via the lungs.

Figure 125 shows representative semilogarithmic plots of regional myocardial (i.e., single crystal) ^{133}Xe washout curves obtained from a patient with coronary

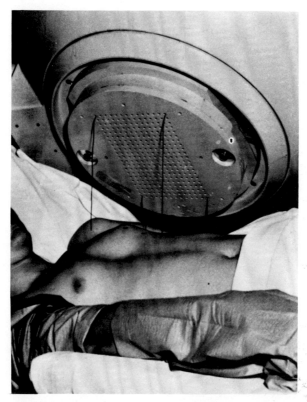

Fig. 122. Photograph of the multiple-crystal scintillation camera and the multichannel collimator used in the studies of regional myocardial blood flow.

artery disease. The data recorded in the first 40 seconds after the peak of each curve were analyzed. The monoexponential equation in (cps/peak cps) = kt was fitted, using least squares, to this initial portion of each curve, and the rate constants of regional ^{133}Xe clearance from the myocardium were calculated. Myocardial nutrient blood flow rates were also calculated by the Kety formula $F_{ml}/100$ g/minute = $k \times \lambda/\rho \times 100$, where k = the rate constant of isotope clearance determined experimentally, $\lambda = 0.72$, the blood/myocardium partition coefficient of ^{133}Xe found by Conn in normal dog heart (66), and ρ is the specific gravity of the tissue (1.05). A confidence limit for each flow measurement, expressed as the standard deviation, may be calculated from the scatter of the original data points about the fitted line. Data from crystals with washout curves showing borderline or low peak counts, or with flow rates with a high standard deviation, occur at the cardiac borders due to motion artifact.

Upon completion of the isotope study, a tracing of the patient's coronary arteriogram (LAO projection) is made after display on a screen at known magnification; the position of the radiopaque–radioactive markers is noted. The computer printout of myocardial perfusion rates is similarly enlarged. The distances between markers (two directions) on both the arteriogram and computer printout are checked to ascertain that the patient's position and the alignment were similar in both the

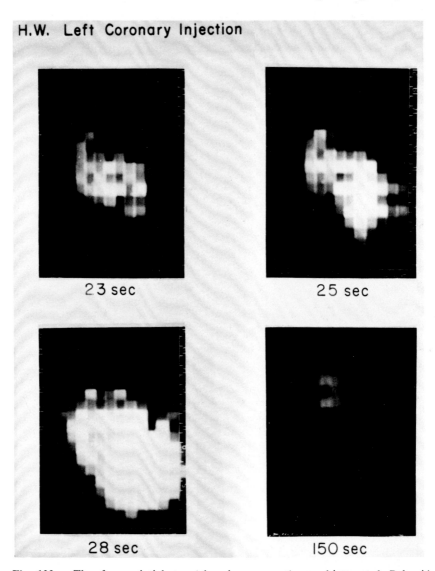

Fig. 123a. The four scintiphotographs shown are 1-second-integrated Polaroid photographs of the cathode-ray oscilliscope of the autofluoroscope. They were obtained during replay of the magnetic tape of the study of patient H. W. in which ^{133}Xe was injected into the left coronary artery. The pictures show isotope location in the myocardium, 1, 3, 5, and 128 seconds after injection which was made at 22 seconds.

radiographic and isotopic parts of the study. The local myocardial perfusion rates are then printed onto the tracing of the arteriogram in appropriate locations.

A myocardial perfusion pattern from a patient with a normal left coronary arteriogram appears in Figure 126. Local myocardial blood flow rates in the figure are expressed in milliliters per 100 g of myocardial tissue per minute; the number in parenthesis beneath each blood flow is the standard deviation of that measure-

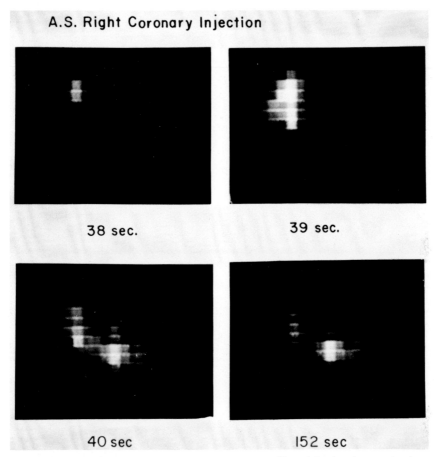

Fig. 123b. Sequential scintiphotographs taken after ^{133}Xe injection into a dominant normal right coronary artery of patient A. S. are shown. A C-shaped radioisotope image was produced.

ment. A small but significant inhomogeneity of local perfusion rates in the left ventricle was observed in this and other patients with normal left coronary arteriograms (35). Mean left ventricular perfusion rate may be obtained by averaging the local flows recorded by all crystals overlying the left ventricle (mean left ventricular myocardial blood flow was 68 ml/100 g/minute in this study). The standard deviation and coefficient of variation of the local perfusion measurements in each study may also be calculated as indices of the heterogeneity of regional myocardial flows in the ventricle. In some studies, the local perfusion rates of the left ventricular subregions supplied by the anterior descending, diagonal, and circumflex branches of the left coronary artery were determined by averaging the perfusion rates of crystals whose area on the perfusion pattern was crossed by the tracing of the artery.

Figure 127 shows the perfusion pattern obtained after ^{133}Xe injection into an arteriographically normal dominant right coronary artery. In right coronary studies different myocardial regions can be distinguished by landmarks provided by the

59	58	197	83	158	118	205	189	152	181	183	137
179	147	172	142	178	176	178	149	150	173	5	172
54	79	156	81	166	105	123	219	114	182	213	132
131	162	149	125	122	134	130	5	5	5	6	5
99	82	167	80	126	155	396	488	327	261	274	106
169	162	165	149	152	117	5	5	5	5	5	107
83	111	150	100	242	399	461	661	615	447	229	108
171	179	121	140	7	5	5	5	5	5	5	94
133	124	149	125	515	461	496	374	355	294	189	116
151	144	150	114	7	7	7	5	5	5	5	95
158	113	172	241	623	447	358	377	173	148	160	117
157	139	128	7	7	7	10	7	7	113	111	89
124	138	292	416	742	325	409	371	112	151	130	109
118	126	14	8	7	7	7	8	64	77	111	72
157	134	442	457	678	318	622	310	113	151	147	146
117	152	13	9	7	8	8	11	68	107	65	81
110	147	379	298	422	272	295	181	113	150	150	108
107	81	9	8	8	8	9	71	108	76	82	85
137	130	401	376	458	249	331	212	132	181	155	123
75	110	9	8	8	9	10	88	71	67	86	63
114	106	304	296	394	218	212	215	127	120	121	100
125	117	9	8	16	11	60	85	56	78	81	68
109	89	117	140	108	117	179	190	107	109	104	106
125	141	154	70	109	88	69	65	66	65	60	94

Fig. 124. A portion of the computer output depicting the peak counts per second recorded by each crystal (top number) and the number of seconds after the start of the study when the peak counts per second in each crystal are presented.

coronary arteriogram. Crystals overlying the right atrial area are located to the patient's right of the main right coronary artery on the perfusion pattern; those crystals between the main right coronary and the A–V node artery overlie the right ventricle, and those to the left of the A–V node artery and the arteriographic "blush" produced by angiographic dye in the interventricular septum overlie that portion of the left ventricle that receives diffusible nutrients from the right coronary artery. In this patient local perfusion rates in the right atrial region averaged 27 ml/100 g/minute, in the right ventricular region, 55 ml/100 g/minute, and in the left ventricular tissue supplied by the right coronary artery, 79 ml/100 g/minute.

Studies of Patients with Normal Coronary Arteriograms

Studies of 17 patients with heart disease but radiographically normal coronary arteries (35) indicated that there are significant regional variations in local myocardial perfusion rates in patients without demonstrable coronary artery disease. (Table 5). The average mean myocardial blood flow rate in the left ventricle, 64.1 ± 13.9 ml/100 g/minute, significantly exceeded that of the right ventricle, 47.8 ± 10.9 ml/100 g/minute, and that of the right atrial region, 33.8 ml/100 g/minute. These results are consistent with similar observations in experimental animals made wtih labeled ions (67) or microspheres (42). Since myocardial oxygen consumption is determined largely by pressure generation in the ventricle (68), the data suggest that the greater pressure work of the left ventricle may require not only a greater muscle mass, but also a larger blood supply per unit tissue than in the right ventricle. In nine patients with normal left arteriograms there was a sig-

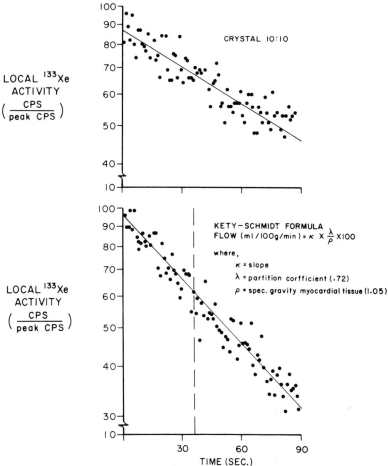

Fig. 125. Two semilogarithmic plots of myocardial ^{133}Xe activity (cps/peak cps versus time) are depicted. These were obtained by two crystals in one patient. The patient had coronary artery disease; one crystal overlay myocardium was supplied by a normal vessel (bottom); the other (top) was over tissue supplied by a narrowed branch of the left coronary.

nificant lack of homogeneity of local perfusion rates in left ventricular myocardium; the coefficient of variation of left ventricular local perfusion rates was 15.8%. That this was true inhomogeneity of flow was indicated by the finding that the between-crystal variance of flows significantly exceeded the individual variance of single-crystal flow measurements. (36)

Preliminary studies have also been completed on two other groups of patients with normal coronary arteriograms. In nine subjects with typical angina pectoris and normal coronary arteriograms, normal regional myocardial perfusion patterns were found (69). Although the measurements were not made at time of exercise or pain, the data tend to exclude occult atheromata or small vessel disease as causal factors of the chest pain in these subjects. Differing myocardial perfusion patterns were observed in patients with myocardial hypertrophy (70). Whereas the left ventricular perfusion patterns were normal in patients with left ventricular hypertrophy and mild hypertension or aortic stenosis, the mean flow per 100 g per min-

Table 5. Summary of Results in Four Groups of Patients [a]

Arteriograms	Flow[b]		Flow[b]		Flow[b]
	No. of Crystals	ml/100 g/ minute	No. of Crystals	ml/100 g/ minute	No. of Crystals	ml/100 g/ minute
	Left atrial region		Left ventricle		coefficients of variation
Normal left (9 patients)	4	41.5 ± 13.7	33	64.1 ± 13.9		15.8
Abnormal left (disease of two branches 12 patients)	7	43.2 ± 9.2	35	48.0 ± 12.0		25.2
	Right atrial region		Right Ventricle		Left Ventricle
Normal right (8 patients)	6	33.6 ± 10.3	13	47.8 ± 10.9	11	69.0 ± 13 (n = 7)
Abnormal right (8 patients)	7	34.3 ± 6.6	18	35.1 ± 4.9	8	40.7 ± 2 (n = 3)

[a] The Table summarizes results obtained in patients with normal and abnormal right and left coronary arteriograms. Mean perfusion data obtained in three cardiac regions in individual studies have been averaged for each group of patients.
[b] Mean flow ± SD of the observation.

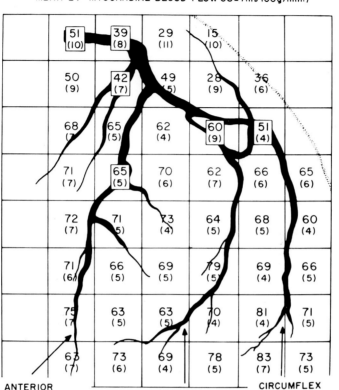

Fig. 126. A myocardial perfusion pattern showing the myocardial blood flow rates (ml/100 g/minute) in different regions of the left ventricle of a patient with a normal left coronary arteriogram is shown. The computer printout has been magnified, aligned, and superimposed on a tracing of the patient's arteriogram (LAO projection) which was obtained during the same study.

ute was significantly subnormal in some normotensive patients with idiopathic cardiomyopathy (Fig. 128). Patients with the latter disease not infrequently had normal or enlarged coronary arteries, rapid transit of angiographic dye, and massive cardiac hypertrophy apparent on ventriculograms. Whether the reduced left ventricular myocardial perfusion resulted from diminished flow or overgrowth of muscle, and whether the low myocardial blood flow per gram contributed to the angina pectoris and/or cardiac failure manifested by some of the patients are problems currently under study.

Studies of Patients with Coronary Atherosclerosis

A spectrum of perfusion patterns has been observed in patients with coronary atherosclerosis (36, 17). In patients with mild disease (<75% obstruction of

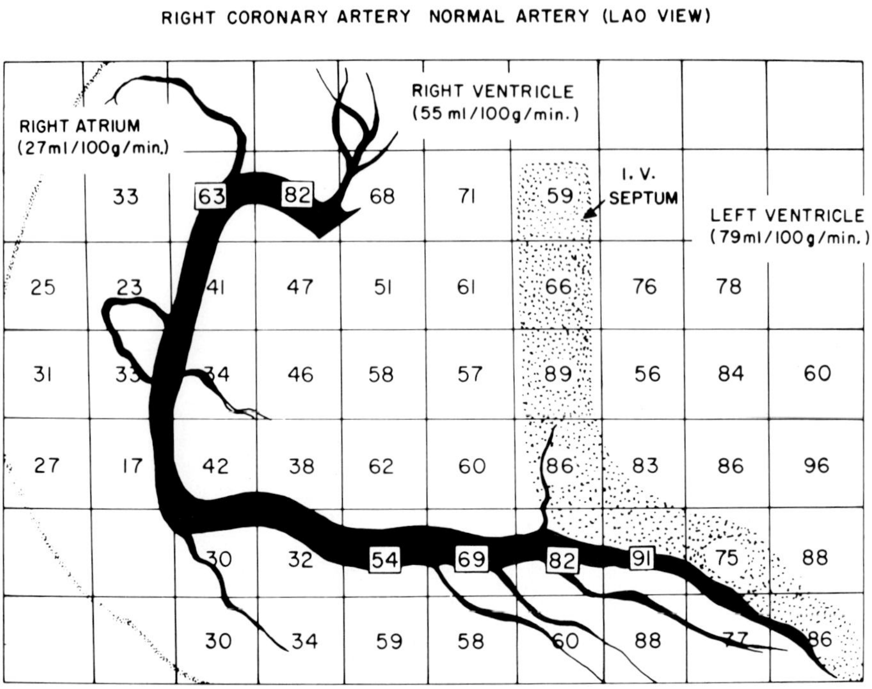

RIGHT CORONARY ARTERY   NORMAL ARTERY (LAO VIEW)

Fig. 127. A computer printout showing local myocardial blood flow/100 g/minute by each crystal has been superimposed onto a tracing of the patient's normal right coronary arteriogram. Different regions of the heart can be distinguished by landmarks provided by the coronary arteriogram.

coronary arteries) the regional myocardial perfusion patterns frequently were not different from those found in patients with normal coronary arteriograms. Preliminary studies have been made in such patients in which regional blood flow measurements have been made at rest and then again during atrial pacing; these have indicated that blood flow in the area beyond coronary lesions may be normal or only slightly diminished at rest, but respond inadequately or not at all to the increased myocardial oxygen consumption of a more rapid heart rate (71).

In other patients with coronary disease and angina, localized regions of reduced myocardial perfusion have been observed at rest which could be related to specific vascular lesions apparent on the arteriogram (36). The study illustrated in Figure 129 was performed 12 weeks after the patient had sustained an anterior myocardial infarction. The coronary arteriogram of the patient showed total occlusion of the left anterior descending artery at its most proximal portion, with only minor disease of the circumflex vessels; the ventriculogram revealed an anterolateral aneurysm. In regions of left ventricle supplied by the diagonal and circumflex branches, the local myocardial perfusion rates averaged 78 ml/100 g/minute, whereas in the aneurysmal region the local perfusion rates were significantly depressed. Figure 130 illustrates the study of a patient with a totally occluded dominant right coronary artery. Myocardial perfusion rates were only slightly reduced in the right atrial region, but were significantly subnormal in the right ventricle. Because the posterior

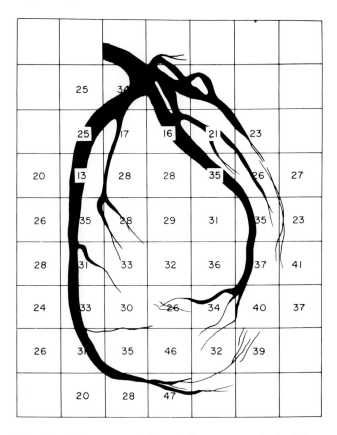

Fig. 128. The myocardial perfusion pattern obtained in a patient with idiopathic myocardial hypertrophy reveals diffusely low flow rates (ml/100 g/minute) throughout the left ventricle. The coronary arteriogram of this patient was normal.

descending artery was totally occluded there was no evidence for local tracer washout from the posterior-inferior left ventricle normally supplied by this artery after a bolus of ^{133}Xe was injected into the right coronary artery.

A study of 20 patients with electrocardiographic evidence of transmural anterior or inferior myocardial infarctions was also performed (72). In patients with anterior infarction, severe obstructive disease of the anterior descending artery was observed on the arteriograms. The mean myocardial perfusion rate in the anterior descending myocardial subregion was 44 ml/100 g/minute, a value significantly less than the mean flow of 62 ml/100 g/minute obeserved in a control group without infarction who had normal left coronary arteriograms. Although the mean perfusion rates in the diagonal and circumflex myocardial subregions were slightly lower in the infarct group than in the control subjects, the differences were not statistically significant. In patients with inferior transmural infarctions, 90% occlusions of the right coronary artery were found. In only one patient was there evidence of ^{133}Xe washout from the inferior left ventricle after the tracer was injected into the right coronary artery. Average mean right ventricular perfusion for the group of 34

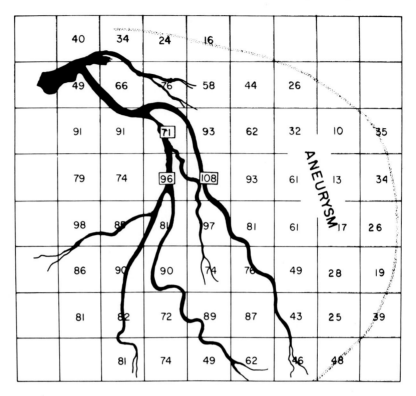

Fig. 129. The myocardial perfusion pattern of a patient with an anterior left ventricular aneurysm.

ml/100 g/minute was significantly less than the value of 48 ml/100 g/minute found in the control group with normal right coronary arteriograms.

Diffuse reductions of myocardial blood flow rates were observed in other patients with coronary atherosclerosis. This is illustrated in Figure 131, which depicts the study of a patient with marked disease of the diagonal and circumflex branches of the left coronary artery and complete occlusion of the anterior descending branch. In a series of 12 patients with >75% obstruction of two or more branches of the left coronary, the average mean resting left ventricular myocardial perfusion rate of 48 ml/100 g/minute was significantly below the value of the 64 ml/100 g/minute found in comparable studies of patients with normal left coronary arteriograms (36). In addition, evidence for increased heterogeneity of myocardial perfusion in coronary heart disease was obtained. The coefficient of variation of local myocardial perfusion rates in the left ventricle in patients with radiographically significant left coronary disease was 25.2%, a value significantly higher than the coefficient of variation of 15.8% found in those with normal arteries (36).

Collateral Blood Flow

This method is capable of detecting whether collateral blood vessels provide adequate nutrient blood flow to a region of myocardium distal to an occluded artery

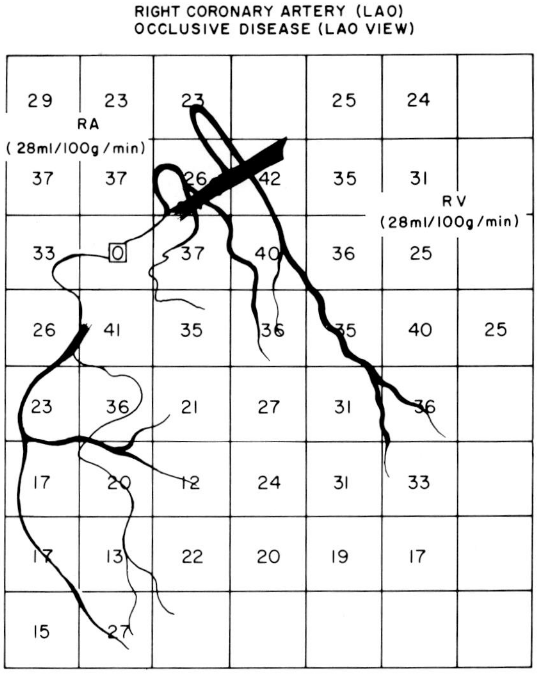

RIGHT CORONARY ARTERY (LAO)
OCCLUSIVE DISEASE (LAO VIEW)

29 RA (28ml/100g/min)	23	23		25	24	
37	37	26	42	35	31 RV (28ml/100g/min)	
33	0	37	40	36	25	
26	41	35	36	35	40	25
23	36	21	27	31	36	
17	20	12	24	31	33	
17	13	22	20	19	17	
15	27					

Fig. 130. Subnormal perfusion rates in right ventricular myocardium and no evidence of tracer washout from left ventricle are apparent on the perfusion pattern of a patient with an occluded dominant right coronary artery.

(36). If sufficient isotope is delivered to an ischemic region by collateral vessels so that washout curves with good counting statistics can be obtained, the rate constants of the local washout curves in that region reflect the adequacy of capillary flow in the area independently of whether the blood flow originates from the injected or other coronary arteries. Figure 132 illustrates the maintenance of right ventricular myocardial perfusion distal to a total right coronary occlusion, albeit at a rate below the average right ventricular perfusion rate found in a control group of patients with normal right coronary arteriograms. Figure 133 illustrates the perfusion pattern obtained when ^{133}Xe was injected into a saphenous vein aortocoronary bypass graft to the distal portion of an occluded left anterior descending artery (71). The number of crystals recording ^{133}Xe-washout curves in this study was related directly and proportionally to the area of myocardial tissue that received diffusible nutrients from the graft via intramyocardial collateral vessels. The calculated perfusion rates reflected the rates of capillary blood flow in each myocardial region regardless of whether some of the blood supply arrived via the graft, via the diseased artery, or via collaterals from other unobstructed coronary branches.

Discussion

The present method for quantitative assessment of capillary perfusion in multiple regions of the myocardium with ^{133}Xe and a scintillation camera is safe, and the

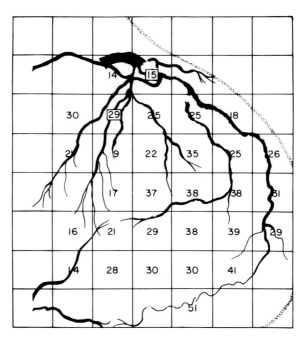

Fig. 131. Diffuse reductions of myocardial perfusion are apparent in the left ventricular myocardium of a patient with marked disease of the diagonal and circumflex branches and complete occlusion of the anterior descending branch of the left coronary artery.

measurements can be performed quickly at the time of coronary arteriography. To date, over 200 patients have been studied in two institutions. With this technique spatial resolution is achieved by collimated individual crystals, an approach which permits analysis of regional/local washout characteristics separately. Earlier approaches at divining regional variations in flow from multiple exponents of a single washout curve provided no specific regional information, and considerable mathematical complexity (73). The indicator washout curves are monitored simultaneously from different areas of the heart, and the rate constants (k) of isotope removal from the myocardium are calculated. The rate constants represent the primary data. Each rate constant is a quantitative expression of the capacity of the local circulation in each myocardial region to eliminate a diffusible substance (and by inference to supply the tissue with diffusible nutrients); each k is directly related to local capillary blood flow by the ratio λ/ρ. The expression of the primary data in terms of blood flow per 100 g/minute, however, involves use of the Kety formula (47) and thus it depends upon: (*a*) assumptions inherent in a monoexponential analysis of the washout curves and (*b*) an assumed partition coefficient (35). The basic approach and observations (i.e., simultaneous measurement of isotope washout from multiple regions of the heart with a multichannel collimator and multiple detectors) retain validity, nevertheless, which are independent of the form of mathematical analysis or of assumptions involved in expressing the data in terms of nutrient blood flow (74).

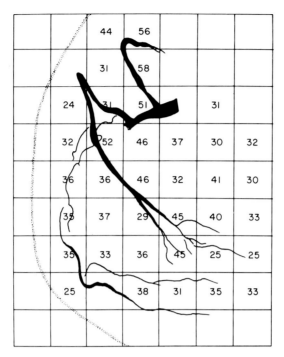

Fig. 132. The myocardial perfusion pattern observed in this patient illustrates that almost normal rates of myocardial perfusion can be maintained in the right ventricle by collateral vessels which arise proximal to a major right coronary occlusion. There is no detectable perfusion in the region of the left ventricle normally supplied by this dominant right coronary vessel.

TWO ARBITRARY CHOICES

In developing this method, several choices were made for theoretical or practical reasons which deserve comment: These include choice of a monoexponential analysis of the data in the first portion of the regional ^{133}Xe washout curves and the use of the blood/myocardium partition coefficient of 0.72 to express the results of local tracer washout in terms of nutrient blood flow per 100 g per minute.

Monoexponential Analysis

In these studies, monoexponential analysis of the data obtained in the first 40 seconds after the peak counts per second of each ^{133}Xe washout was performed, although washout data were recorded for 4 to 7 minutes. This analysis was chosen for several reasons: (*a*) the initial portions of the curves are clearly linear when plotted semilogarithmically against time (35); (*b*) the initial slope of the curve is not affected by isotope arrival in chest wall or lung behind the heart and is probably least affected by the content of nonmuscular tissue (fibrosis or fat) in the

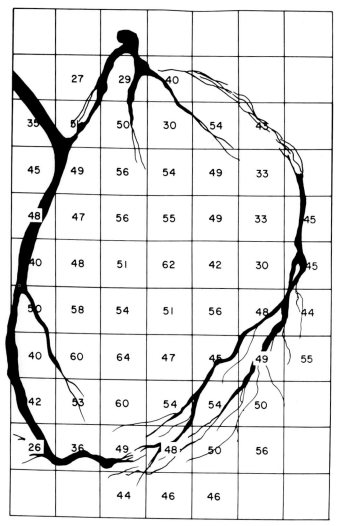

Fig. 133. This is the myocardial perfusion pattern observed when ^{133}Xe was injected into a graft into the left anterior descending artery distal to a >90% obstruction of the vessel. The study indicates that a significant area of the anterior left ventricle received diffusible nutrients from this vessel.

field of view of each crystal (39); (c) counting statistics deteriorate as one moves down the washout curve beyond 40 to 60 seconds; (d) most importantly, four studies indicate that in experimental animals there was a close correspondence between directly measured (flow meter) coronary artery flow per gram tissue and myocardial blood flow rates calculated by monoexponential analysis of the initial segment of ^{133}Xe or ^{85}Kr (single crystal) myocardial washout curves (37–39, 51). Figure 134, taken from Shaw, Pitt, and Friesinger (51), shows the correspondence observed between ^{133}Xe and flow meter myocardial blood flows in experiments in 22 perfused normal dog hearts in which a monoexponential analysis of the initial portion of the isotope washout curves was performed.

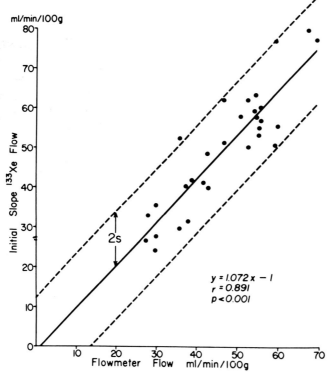

Fig. 134. Comparison of myocardial blood flow per 100 g perfused myocardium in the isolated heart as calculated from the initial portion of the precordial disappearance curve of ^{133}Xe on the ordinate with simultaneously measured flow meter flow per 100 g on the abscissa [From Shaw et al. (38) by permission of the Editors.]

Height–area analysis (75) is difficult to use in myocardial perfusion studies for several reasons: (*a*) there is uncertainty where to terminate such analysis (76); (*b*) flow data from such analysis of ^{133}Xe data did not correlate with simultaneous directly measured flows in dogs (39); and (*c*) observations that after 40 seconds ^{133}Xe accumulated in lung behind the heart as it washed out of myocardium (35) suggested that such a buildup of intrapulmonary counts behind the heart might seriously distort height–area analysis.

Several groups of investigators have performed two- or three-compartment analysis of ^{133}Xe myocardial washout curves (40, 73, 77–79). However, myocardial flow per gram values calculated with this type of analysis did not agree as well with directly measured coronary flow per gram as did values obtained from monoexponential analysis of the same data (39, 61). In addition, in none of the published studies in which two- or three-compartment analysis has been used (40, 73, 77–79) is it clear to what extent the deviation of the tracer curve from a single exponential, observed after 40 to 90 seconds, may have resulted from washout of tracer from cardiac fibrosis and fat, from arrival of tracer in chest wall overlying the heart, or from arrival of tracer in lung tissue behind the heart or elsewhere within the field of view of the detector.

Partition Coefficient

In expressing the results of the studies in terms of nutrient myocardial blood flow, each of the rate constants of local isotope clearance from the myocardium was multiplied by a factor (λ/ρ). The value used for λ in the calculation was 0.72, that found by Conn in static studies of normal canine myocardium (69). The expression of the primary data, the rate constants of tracer clearance, in terms of blood flow must be interpreted with caution, therefore, to the extent that the values depend upon an assumed λ. Data concerning the λ of human cardiac muscle, or for the λ which might exist in diseased cardiac tissue are not generally available. Choice of the λ of 0.72 was made, in part, because the flowmeter studies (Fig. 134) showed good correlation with myocardial flow calculated from ^{133}Xe washout curves using this value.*

The partition coefficient for ^{133}Xe in fat is 8. The possibility that ^{133}Xe washout curves might be altered in different regions of the heart because of disease-induced changes in local λ due to differing concentration of muscle, fat, and other tissue components has not been approached experimentally. The selection both of initial slope analysis and a λ of 0.72 was made also to minimize any effects of cardiac fat upon the measurements of myocardial perfusion. This may occur since: (a) loading of cardiac fat after slug injections of inert gas is slow because adipose blood flow is a small fraction of myocardial blood flow (41); (b) ^{133}Xe has not equilibrated with fat by diffusion at 30 to 40 seconds (39); and (c) persistence of isotope in cardiac fat influences the tail of ^{133}Xe washout curves to a much greater extent than it influences the more rapidly changing initial portions of the curve (39). In addition, radioautographs by Shaw et al. (61) and scintiphotographs of isotope distribution in human hearts (74) have shown ^{133}Xe to be homogeneously distributed throughout myocardium during the first 40 seconds of washout; localization in epicardial fat was not apparent until after 3 minutes (when measurements of tracer washout by the present technique would be complete).

The use of antipyrine ($\lambda = 1$) or labeled H_2O would obviate the possibility that cardiac fat distorts tracer washout independently of blood flow and could be used in this type of study if the problems of recirculation and/or availability of a cyclotron could be overcome. Pertinent to the present technique, however, are data from a study using isolated dog hearts by Bassingthwaighte et al. (39). These workers compared flow per gram measurements made in the same hearts using ^{133}Xe and ^{125}iodo-antipyrine. The regression equation relating the two flows was $F_{Xe} = y^0 + bF_{Ap}$. Using initial slope analysis for the ^{133}Xe curves the whole heart values were $y_0 = -13$, $b = 1.04$, $SD = 17$, $r = 0.99$; values obtained at the cardiac apex were $y_0 = -9$, $b = 1.04$, $SD = 19$, $r = 0.99$ (Fig. 135). Even though this relationship may be fortuitous, it suggests that myocardial blood flow estimates using initial slope monoexponential analysis and ^{133}Xe compare well with estimates made under the same coronary flow conditions with a fat-insoluble indicator such as antipyrine. It seems unlikely, therefore, that epicardial fat distorted the estimates of regional myocardial perfusion rates found in the studies of normal human heart (35, 36). Whereas distortion of tracer elimination by fatty replacement within

* Ladefoged (51) has shown that the λ for ^{133}Xe in kidney tissue is influenced by hematocrit; the correction factor developed by these workers for hematocrit was not applied in the present studies, since most of the patients had a normal hematocrit.

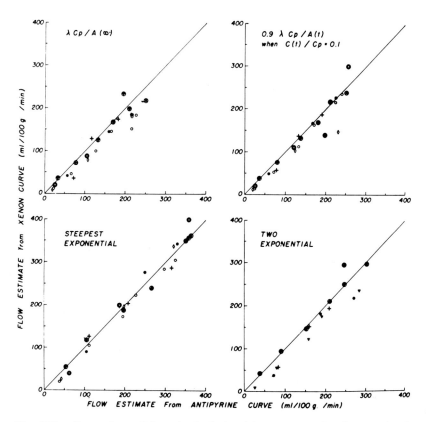

Fig. 135. Comparison of flows in the isolated heart preparation from antipyrine and xenon curves by four methods. [From Bassingthwaighte et al. (16), by permission of the Editors.] The circled symbols indicate estimates from curves obtained by collimation on the apex. Each xenon curve was recorded 3 to 8 minutes after the antipyrine curve with which it was paired at constant perfusion rate. Regression equations are given in the text. The graph in the lower left shows the relationship between ^{133}Xe flow/100 g-minute calculated by monoexponential analysis of the initial slope of the curves with that obtained with labeled antipyrine.

myocardial tissue cannot be completely excluded in pathological conditions, the published experimental data (38, 39, 51) would suggest that the analytic technique employed in the present studies tends to minimize any such effects. Furthermore, a recent modification of the technique, whereby the patient's regional myocardial perfusion is studied before and during atrial pacing (71), obviates this problem, since changes in local perfusion in response to increased heart rate occur independently of the composition of the tissue.

Advantages of Method

As it is currently used, the present method has several advantages. Studies can be rapidly and safely performed at the time of coronary arteriography. The data provide quantitative estimates of regional myocardial perfusion from measurements carried out simultaneously in multiple regions of the heart; because the myocardium supplied by each artery is subdivided into separate regions from which individual ^{133}Xe washout curves are obtained, the uncertainties of attempting to resolve single multiexponential washout curves which result from inhomogeneous myocardial flow in coronary disease are avoided. By use of radioactive–radiopaque landmarks present during both isotopic and radiographic assessments of the coronary arteries, areas of abnormal myocardial perfusion can be related to specific vascular lesions. Injection of ^{133}Xe selectively into the coronary artery minimizes effects of tracer in the chest wall and obviates need for a thoracotomy or direct needle injection; rapid pulmonary elimination minimizes recirculation. The current method of data analysis also gives a quantitative statistical estimate of the accuracy of each individual measurement (important in comparing different regional flow values). Because myocardial nutrient perfusion is measured, collateral blood flow can be assessed. The number and position of the detectors recording tracer removal after coronary injection reflect the size and location of areas of myocardium which receive diffusible nutrients from that artery. Because the measurement is dynamic and brief, multiple studies (e.g., before and during atrial pacing) can be performed in order to measure regional flow responses to increased oxygen requirements of greater myocardial work.

Limitations of Method

The technique, in its current state of development, has four types of limitations which are related to: (*1*) the indicator, ^{133}Xe, (*2*) the technology of the instrumentation, (*3*) cardiac motion, and (*4*) the fact that each detector views isotope removal from all tissue within its field of view.

Limitation *1*: The problem of the fat solubility of ^{133}Xe as it relates to the blood/myocardium has been discussed above. Another potential problem with ^{133}Xe relates to the fact that ^{133}Xe is administered as a bolus injection to avoid the accumulation of tracer in precordium which would occur if it were given by inhalation. Studies in which single H_2 washout curves were monitored after the myocardium was loaded with H_2 by inhalation or injection have suggested that areas of reduced myocardial perfusion might not be fully loaded with inert gas after a bolus injection. In order to detect washout of H_2 from areas of reduced flow, Klocke et al. (41, 80, 81) monitored coronary artery–coronary sinus H_2 concentration differences for a long period of time. The necessity for prolonged monitoring is avoided in the present method by use of the multichannel collimator and multiple detectors to physically separate regions of the myocardium and to identify areas of reduced perfusion by the changes in the rate of ^{133}Xe clearance from these discrete regions. It is quite possible that with a bolus injection of ^{133}Xe little isotope might reach an area of reduced perfusion beyond a coronary occlusion; however, if such an event occurs, the amount which was delivered is known from the printout of the peak counts and time after injection recorded for each crystal (Fig. 12). If too few counts arrive in a region, the standard deviation of the local tracer washout curve is high and the computer is unable to apply a monoexponential equation to the data (e.g., random scatter of data in areas of scar). In general,

Limitations of Method

however, washout curves with an acceptable accuracy have been observed in regions with flow per gram as low as 20 to 30 ml/100 g/minute.

The low photon energy of ^{133}Xe, 80 keV, results in extensive narrow angle scatter from Compton interactions in the chest wall, and in collimator septa. This physical factor reduces very substantially the spatial resolution with which data can be recorded. Another effect of the low energy photon is tissue absorption; the half value level for tissue is less than 3 cm. Thus the efficiency for counting the anterior wall of the myocardium is better than 50%, but efficiency for the posterior wall is less than 15%. Another radioisotope of xenon, ^{127}Xe, with a photon energy of 209 keV and a half value layer of 5 cm will soon be available, providing better resolution, and higher counting efficiency, particularly necessary for studying the myocardium lying deepest in the chest.

Limitation 2: The main limitations of the method directly related to the instrumentation are as follows: (*a*) the current dead-time which limits maximum counting rate and therefore the measurement accuracy. If more counts could be obtained per counting interval in each region during tracer washout, the standard deviation would be smaller and the ability to detect intramyocardial regional flow variations would be increased; (*b*) the intrinsic resolution of the multiple-crystal instrument is limited by the size of the crystals (1.1 × 1.1 cm). The radius of the field of view of each crystal at distances of 5 to 10 cm could be reduced by using a longer collimator. This in turn would reduce the area of myocardial wall viewed by each crystal and also reduce the degree to which overlap of field of view occurs at the back wall of the heart (a phenomenon which reduces measurement accuracy in that region) (35).

Limitation 3: Cardiac motion obviously introduces imprecision into the measurements of isotope removal from small myocardial regions. Currently this error is included in the standard deviation which is calculated from the scatter of the observed data points about the least squares slope of the initial portion of each ^{133}Xe washout curve. A more sensitive instrument might facilitate more rapid framing of data in real time so that counts obtained at a given point in different cardiac cycles (EKG) could be recorded to reduce that component of measurement error due to motion.

Limitation 4: The fourth type of limitation of the current technique relates to the fact that ^{133}Xe washout from all tissue within the field of view of each crystal is monitored. Because of the spherical nature of the heart, two or more myocardial surfaces of the heart may obviously be viewed by the same crystal in certain projections. For this reason most of the present studies of the left coronary artery have been performed in the left anterior oblique projection which displays the areas supplied by the three left coronary branches to greatest advantage. With a more efficient instrument, thicker collimation, and a better tracer isotope, it may someday be possible to make regional flow measurements from multiple views in the same patient selecting projections that display the myocardial regions distal to coronary vascular lesions with minimal overlap of other myocardial tissue supplied by the same vessel.

Because each scintillation detector monitors ^{133}Xe removal from all the radioactive tissue within its field of view, the question arises whether flow differences between endocardium and epicardium might be detected. Although subendocardial infarction in man has been reported (82) there is no evidence concerning transmural variation in myocardial perfusion in human disease. Studies in animals with diffusible indicators have indicated that at rest the subendocardial/subepicardial flow ratio was slightly greater than 1 (83–86). Studies during induced reductions of coronary flow using labeled 15 μ microspheres, however, showed modest reductions of the endocardial/epicardial ratio of the labeled spheres (87). The observations in dog hearts perfused at known flows from 1.3 to 4.8 ml/g/minute show that the ratio of deposits in inner to outer left ventricle was 4.47 with 50 μ spheres, 3.48 with 35 μ spheres, 5.97 with 10 μ spheres, and 2.67 with <10 μ spheres (86), indicating that rheological streaming at intramyocardial arterial branch points influences sphere distribution across the heart wall and probably invalidates use of spheres greater than 10 μ for making such measurements. (Since the inner/outer ratio was 2.67 with 10 μ spheres, 1.76 with iodo-antipyrine, and 1.32 with Rb, even 10 μ spheres may not be optimal.) Nevertheless, the accumulated data suggest that transmural perfusion gradients do not occur at rest in normal hearts, but may result during myocardial ischemia.

If one assumes that there might develop two flow compartments across the heart wall during reduced coronary artery flow, the data observed by a single detector monitoring ^{133}Xe washout from this region would show a two component curve; the equations indicate that what is observed is in fact the sum of two exponential washout curves. The initial slope of the observed curve, however, is the derivative at time zero of the observed curve (Equation 4). Thus an initial slope monoexponential analysis of a hypothetical biexponential washout of ^{133}Xe from myocardium with subendocardial ischemia would yield a perfusion value that would represent a weighted average of the two rate constants present in the region under study.

In order to test whether a monoexponential initial slope analysis of biexponential washout curves would yield directionally incorrect or grossly erroneous values for weighted mean flow across the myocardial wall, the mean transmural myocardial perfusion data obtained by Buckberg et al. (58) which showed varying endo/epi(10 μ microspheres) ratios was analyzed according to the equations above. Weighted-average myocardial flows calculated in this way differed by $<5\%$ from those calculated using the microsphere data (58, Table 1) and never were directionally opposite.

Potential Uses

The present approach to the quantitative assessment of regional myocardial perfusion in man with ^{133}Xe and a scintillation camera has enabled the patterns of regional myocardial blood flow present in intact patients with normal coronary arteriograms to be measured for the first time (34, 35). Differing patterns, which showed increased heterogeneity of local myocardial perfusion rates, have also been identified in patients with ischemic heart disease and with various forms of cardiac hypertrophy (34–35). Despite its limitations in the current stage of development, the method would appear to have several immediate clinical and investigative applications if used together with high quality selective coronary arteriography: (1) localization of ischemia or functionally avascular areas of the myocardium, (2) study of the effectiveness of collateral blood supply to areas beyond coronary artery occlusive lesions, (3) study of diseases which affect the coronary microcirculation,

(*4*) investigation of the effects of drugs upon ischemic myocardium, (*5*) evaluation of the effectiveness of surgical procedures to revascularize ischemic heart tissue, and (*6*) study of the relationships between myocardial blood flow and cardiac performance in coronary and hypertensive heart disease.

REFERENCES

1. *Vital Statistics in the United States,* Volume II, Part A, Mortality, 1968, U.S. Department of Health, Education and Welfare, Public Health Service, Rockville, Maryland, 1972.
2. R. Favolaro: Direct and Indirect Coronary Surgery, *Circulation* **46**: 1197–1208, 1972.
3. J. C. Manley, and W. D. Johnson: Effects of Surgery on Angina (Pre and Post-Infarction) and Myocardial Function (Failure), *Circulation* **46**: 1208–1222, 1972.
4. F. M. Sones, Jr., and E. K. Shirley: Cine Coronary Arteriography, *Mod. Concepts Cardiovasc. Dis* **31**: 735–738, 1962.
5. H. G. Kemp, W. C. Elliott, and R. Gorlin: The Anginal Syndrome with Normal Coronary Arteriography, *Trans. Assoc. Am. Physicians* **80**: 59–70, 1967.
6. L. S. Cohen, W. C. Elliott, M. D. Klein, and R. Gorlin: Coronary Heart Disease: Clinical Cinearteriographic and Metabolic Correlations, *Amer. J. Cardiol.* **17**: 153–169, 1965.
7. C. K. Friedberg: Some Comments and Reflections on Changing Interests and New Developments in Angina Pectoris, *Circulation* **46**: 1037–1038, 1972.
8. R. Gorlin: Pathophysiology of Cardiac Pain, *Circulation* **32**: 138–149, 1965.
9. L. Donato, G. Bartolomei, and R. Giordani: Evaluation of Myocardial Blood Perfusion in Man with Radioactive Potassium or Rubidium and Precordial Counting, *Circulation* **29**: 195–204, 1964.
10. L. Donato, G. Bartolomei, G. Federight, and G. Torreggiani: Measurement of Coronary Blood Flow by External Counting with Radioactive Rubidium, *Circulation* **33**: 708–719, 1966.
11. N. Goldschlager, K. Ravens, G. Leb, C. Cowan, and R. J. Bing: Studies on Effective Capillary Blood Flow of the Human and Canine Heart after Acute Myocardial Infarction, *Circulation (Suppl. IV)*, **40**: 145–155, 1969.
12. S. B. Knoebel, P. L. McHenry, L. Stein, and A. Sonel: Myocardial Blood Flow in Man as Measured by a Coincidence Counting System and a Single Bolus Injection of $Rb^{84}Cl$, *Circulation* **36**: 187–192, 1967.
13. L. A. Sapirstein: Fractionation of the Cardiac Output in Rats with Isotopic Potassium, *Circulation Res.*, **4**: 689–692, 1956.
14. L. Donato: Quantitative Radiocardiography and Myocardial Blood Flow Measurements with Radioisotopes, in *Dynamic Studies with Radioisotopes in Medicine,* IAEA, Vienna p. 645, 1971.
15. N. D. Poe: Comparative Myocardial Uptake and Clearance Characteristics of Potassium and Cesium, *J. Nucl. Med.* **13**: 557–560, 1972.
16. W. J. Lorenz, P. Schmidlin, H. Kampmann, H. Ostertag, W. E. Adam, H. Arnold, and W. Maier-Borst: Comparative Investigations with the Anger Scintillation Camera and the Digital Autofluoroscope, in *Medical Radioisotope Scintigraphy,* Vol. I, IAEA, Vienna p. 135, 1969.
17. W. J. MacIntyre, and T. S. Houser: Experimental Derivations of Point-spread Functions from Line-spread Recordings, *J. Nucl. Med.* **12**: 379, 1971.
18. W. J., MacIntyre, R. E. Botti, Y. Ishii, and T. S. Houser: Localization of Heart Chambers by Temporal and Spatial Analysis for Regional Myocardial Blood Flow Measurements, Proceedings and Symposium on Sharing of Computer Programs and Technology in Nuclear Medicine, Springfield, Virginia, 1972, p. 253. CONF - 720430.

19. T. Nagai, T. A. Iinuma, and S. Koda: Computer-Focusing for Area Scans, *J. Nucl. Med.* **9:** 507–516, 1968.
20. W. J. MacIntyre, Y. Ishii, W. H. Pritchard, and R. W. Eckstein: Measurement of Total Myocardial Blood Flow with ^{43}K and the Scintillation Camera. Myocardial Blood Flow in Man. Proceedings of the Pisa Symposium, June 1971, Minerva Medica, Torino, 207–218, 1972.
21. R. E. Botti, W. J. MacIntyre, and W. H. Pritchard: Identification of Ishemic Area of Left Ventricle by Visualization of ^{43}K Myocardial Deposition, *Circulation* **47:** 486–492, 1973.
22. R. E. Botti, and W. J. MacIntyre: Regional Myocardial Blood Distribution Using Intracoronary ^{43}K, *Clin. Res.* **20:** 767, 1972.
23. M. Endo, T. Yamazaki, K. Soki, H. Hiroo, A. Tomio, T. Tsuguhito, and S. Sakakibara: The Direct Diagnosis of Human Myocardial Ischemia Using ^{131}I-MAA via the Selective Coronary Catheter, *Am. Heart J.* **80:** 498–506, 1970.
24. W. L. Ashburn, E. Braunwald, A. L. Simon, K. L. Peterson, and J. H. Gault: Myocardial Perfusion Imaging with Radioactive-Labeled Particles Injected Directly into the Coronary Circulation of Patients with Coronary Artery Disease, *Circulation* **44:** 851–865, 1971.
25. C. Jansen, M. Judkins, G. Grames, M. Gander, and R. Adams: Myocardial Perfusion Color Scintigraphy with Radioactive MAA in 400 Patients Undergoing Coronary Angiography, *Radiology* **109:** 369–380, 1973.
26. J. L. Quinn, III, M. Serratto, and P. K. Kezdi: Coronary Artery Bed Photoscanning Using Radioiodine Albumin Macroaggregates (RAMA), *J. Nucl. Med.* **7:** 107–113, 1966.
27. H. R. Schelbert, W. L. Ashburn, J. W. Covell, A. L. Simon, E. Braunwald, and J. Ross, Jr.: Feasibility and Hazards of the Intracoronary Injection of Radioactive Serum Albumin Macroaggregates for External Myocardial Perfusion Imaging, *Invest. Radiol.* **6:** 379–387, 1971.
28. N. D. Poe: The Effects of Coronary Arterial Injection of Radioalbumin Macroaggregates on Coronary Hemodynamics and Myocardial Functions, *J. Nucl. Med.* **12:** 724–731, 1971.
29. D. A. Weller, R. J. Adolph, H. N. Wellman, R. G. Carroll, and O. Kim: Myocardial Perfusion Scintigraphy after Intracoronary Injection of ^{99m}Tc-Labeled Human Albumin Microspheres, *Circulation* **46:** 963–975, 1972.
30. L. G. Martin, J. H. Larose, R. G. Sybers, D. H. Tyras, and P. N. Symbas: Myocardial Perfusion Imaging with ^{99m}Tc-Albumin Microspheres, *Radiology* **107:** 367–370, 1973.
31. C. Jansen, M. Judkins, and G. Grames: Myocardial Perfusion Scintigraphy with Radioactive MAA, 13th Symposium on Nuclear Medicine, Oak Ridge, Tennessee, 1972. CONF-711101.
32. P. J. Cannon, J. I. Haft, and P. M. Johnson: Visual Assessment of Regional Myocardial Perfusion Using Radioactive Xenon and Scintillation Photography. *Circulation* **40:** 277, 1969.
33. P. J. Cannon, R. B. Dell, E. M. Dwyer, and P. M. Johnson: A Method for Quantitative Assessment of Regional Myocardial Perfusion in the Intact Animal, *Circulation* **40:** III–56, 1969.
34. P. J. Cannon, R. B. Dell, and E. M. Dwyer, Jr.: Regional Myocardial Perfusion in Man, *J. Clin. Invest.* **49:** 6a, 1970.
35. P. J. Cannon, R. B. Dell, and E. M. Dwyer, Jr.: Measurement of Regional Myocardial Perfusion in Man with ^{133}Xenon and a Scintillation Camera, *J. Clin. Invest.* **51:** 964, 1972.
36. P. J. Cannon, R. B. Dell, and E. M. Dwyer, Jr.: Regional Myocardial Perfusion Rates in Patients with Coronary Artery Disease, *J. Clin. Invest.* **51:** 978, 1972.
37. J. A. Herd, M. Hollenberg, G. D. Thorburn, H. H. Kopald, and A. C. Barger: Myocardial Blood Flow Determined with Krypton^{85} in Unanesthetized Dogs, *Amer. J. Physiol.* **203:** 122, 1962.
38. R. S. Ross, K. Ueda, P. R. Lichtlen, and J. R. Rees: Measurement of Myocardial Blood Flow in Animals and Man by Selective Injection of Radioactive Inert Gas into the Coronary Arteries, *Circulation Res.* **51:** 28, 1964.

References

39. J. B. Bassingthwaighte, T. Strandell, and D. E. Donald: Estimation of Coronary Blood Flow by Washout of Diffusible Indicators, *Circulation Res.* **23:** 259, 1968.
40. M. D. Klein, L. S. Cohen, and R. Gorlin: Krypton[85] Myocardial Blood Flow: Precordial Scintillation versus Coronary Sinus Sampling, *Am. J. Physiol.* **209:** 705, 1965.
41. F. J. Klocke, R. C. Koberstein, D. E. Pittman, I. L. Bunnell, D. G. Greene, and R. D. Rosing: Effects of Heterogeneous Myocardial Perfusion on Coronary Venous H_2 Desaturation Curves and Calculations of Coronary Flow, *J. Clin Invest.* **47:** 2711, 1968.
42. R. J. Domenech, J. I. E. Hoffman, M. I. M. Noble, K. B. Saunders, J. R. Hensen, and S. Subijanto: Total and Regional Coronary Blood Flow Measured by Radioactive Microspheres in Conscious and Anesthetized Dogs, *Circulation Res.* **25:** 581, 1969.
43. W. Hollander, I. M. Madoff, and A. V. Chobanian: Local Myocardial Blood Flow as Indicated by the Disappearance of NaI[131] From Heart Muscle: Studies at Rest, During Exercise and Following Nitrite Administration, *J. Pharmacol. Exp. Therap.* **139:** 53, 1963.
44. J. J. Sullivan, W. J. Taylor, W. C. Elliott, and R. Gorlin: Regional Myocardial Blood Flow, *J. Clin. Invest.* **46:** 1402, 1967.
45. W. L. Ashburn, E. Braunwald, A. L. Simon, K. L. Peterson, and J. H. Gault: Myocardial Perfusion Imaging with Radioactive Labelled Particles Injected Directly into the Coronary Circulation of Patients with Coronary Artery Disease, *Circulation* **44:** 851, 1971.
46. D. A. Weller, R. J. Adolf, H. N. Wellman, R. G. Carroll, and O. Kim: Myocardial Perfusion Scintigraphy after Intracoronary Injection ^{99m}Tc-Labelled Human Albumin Microspheres: Toxicity and Efficacy for Detecting Myocardial Infarction in Dogs: Preliminary Results in Man, *Circulation* **46:** 963, 1972.
47. W. J. MacIntyre, Y. Ishii, W. H. Pritchard, and R. W. Eckstein. The Measurement of Myocardial Blood Flow with [43]Potassium and the Scintillation Camera, in *Myocardial Blood Flow in Man: Methods and Significance in Myocardial Disease*, L. Donato and A. Maseri, Eds., S. Karger, Basel, Switzerland, 1971.
48. A. Cohen, E. J. Zaleski, H. Baleiron, T. B. Stock, C. Chiba, and R. J. Bing: Measurement Method—a Critical Analysis, *Am. J. Cardiol.* **19:** 556, 1967.
49. T. Yipintsoi, R. G. Tancredi, D. Richmond, and J. B. Bassingthwaighte: Myocardial Extractions of Sucrose, Glucose and Potassium, in *Capillary Permeability*, C. Crone and N. A. Lassen, Eds., Academic Press, New York, 1970. pp. 60–80.
50. G. A. Langer. Ion Fluxes in Cardiac Excitation and Contraction and Their Relation to Myocardial Contractility, *Physiol. Rev.* **48:** 708, 1968.
51. J. Ladefoged: Measurement of the Renal Blood Flow in Man with the [133]Xenon Washout Technique, *Scand. J. Clin. Lab. Invest.* **18:** 299, 1966.
52. N. A. Lassen, and A. Klee: Cerebral Blood Flow Determined by Saturation and Desaturation with Krypton[85]; Evaluation of the Inert Gas Method of Kety and Schmidt, *Circulation Res.* **16:** 26, 1965.
53. N. A. Lassen, J. Lindgjerg, and O. Munck: Measurement of Blood Flow through Skeletal Muscle by Intramuscular Injection of Xenon 133, *Lancet* **1964:-I:** 686.
54. C. A. Chidsey, III, H. W. Fritts, Jr., A. Hardewig, D. W. Richards, and A. Cournand, Fate of Radioactive Krypton (Kr[85]) Introduced Intravenously in Man, *J. Appl. Physiol.* **14:** 63, 1959.
55. N. A. Lassen: Assessment of Tissue Radiation Dose in Clinical Use of Radioactive Inert Gases with Examples of Absorbed Doses from 3-H, 85-Kr, 133-Xe, *Minerva Nucl.* **8:** 211, 1964.
56. S. S. Kety: Theory and Application of Exchange of Inert Gas at Lungs and Tissues, *Pharmacol. Rev.* **3:** 1, 1951.
57. S. S. Kety: I. Blood-Tissue Exchange Methods. Theory of Blood-Tissue Exchange and its Application to Measurement of Blood Flow, *Methods Med. Res.* **8:** 223, 1960.
58. G. D. Buckberg, D. E. Fixler, J. P. Archie, and J. I. E. Hoffman: Experimental Subendocardial Ischemia in Dogs with Normal Coronary Arteries, *Circulation Res.* **30:** 67, 1972.

59. T. Yipintsoi, T. H. Knapp, and J. B. Bassingthwaighte. Countercurrent Exchange of Labelled Water in Canine Myocardium, *Fed. Proceedings* (Abstr.) **28**: 645, 1969.
60. T. Yipintsoi, and J. B. Bassingthwaighte: Circulatory Transport of Iodoantipyrine and Water in the Isolated Dog Heart, *Circulation Res.* **27**: 461, 1970.
61. D. J. Shaw, A. Pitt, and G. C. Friesinger: Autoradiographic Study of the ^{133}Xenon Disappearance Method for Measurement of Myocardial Blood Flow, *Cardiovasc. Res.* **6**: 268, 1971.
62. H. O. Anger: Scintillation Camera with Multichannel Collimators, *J. Nucl. Med.* **5**: 515, 1964.
63. M. A. Bender: *Recent Advances in Nuclear Medicine,* M. Croll and O. O. Brady, Eds., Appleton-Century Crofts, New York, 1966.
64. R. Grenier, and J. V. DiRocco: Performance Characteristics of the Digital Autofluoroscope, *IEEE Trans. Nucl. Sci.* **15**: 366, 1968.
65. M. P. Judkins: Selective Coronary Arteriography by the Percutaneous Femoral Technique, *Radiology* **87**: 815, 1967.
66. H. L. Conn, Jr.: Equilibrium Distribution of Radio Xenon in Tissue: Xenon-Hemoglobin Association Curve, *J. Appl. Physiol.* **16**: 1065, 1961.
67. W. D. Love, G. E. Burch: Differences in the Rate of Rb^{86} Uptake by Several Regions of the Myocardium of Control Dogs and Dogs Receiving 1-Norepinephrine or Pitressin, *J. Clin. Invest.* **36**: 479, 1957
68. E. Braunwald: The Determinants of Oxygen Consumption, *The Physiologist* **12**: 65, 1969.
69. E. M. Dwyer, Jr., R. B. Dell, and P. J. Cannon: Regional Myocardial Blood Flow in Patients with Angina and Normal Coronary Arteries, *Circulation* **46**: II-6, 1972.
70. M. Weiss, D. H. Schmidt, C. Jaffe, W. Casarella, K. Ellis, and P. J. Cannon: Regional Myocardial Blood Flow in Patients Idiopathic Myocardial Hypertrophy, *Circulation* **46**: II-235, 1972.
71. D. H. Schmidt, M. B. Weiss, W. Casarella, K. Ellis, and P. J. Cannon, Personal Communication.
72. E. M. Dwyer, Jr., R. B. Dell, and P. J. Cannon: Regional Myocardial Blood Flow in Transmural Myocardial Infarction, *Circulation* **44**: II-125, 1971.
73. R. W. Parkey, S. E. Lewis, E M. Stokely, and F. J. Bonte: Compartmental Analysis of ^{133}Xenon Regional Myocardial Blood Flow Curve, *Radiology* **104**: 425, 1972.
74. A. Maseri, and P. Mancini: Regional Myocardial Blood Flow Distribution in Man by ^{133}Xenon and Radioactive Microspheres Using an Automated Scintillation Camera, in *Myocardial Blood Flow in Man: Methods and Significance in Myocardial Disease,* L. Donato, and A. Maseri, Eds., S. Karger, Basel, S. Switzerland, 1971.
75. K. L. Zierler: Equations for Measuring Blood Flow by External Monitoring of Radioisotopes, *Circulation Res.* **16**: 309, 1965.
76. J. B. Bassingthwaighte: Blood Flow and Diffusion through Mammalian Organs, *Science (Washington)* **167**: 1347, 1970.
77. L. O. Horwitz, R. Gorlin, and W. J. Taylor: Effects of Nitroglycerine on Regional Myocardial Blood Flow in Coronary Artery Disease, *J. Clin. Invest.* **50**: 1578, 1971.
78. E. Linder: Measurements of Normal and Collateral Coronary Blood Flow by Coronary Arterial and Intramyocardial Injection of Krypton 85 and Xenon 133, *Acta Physiol. Scand., Suppl. 272* **68**: 5, 1966.
79. S. C. Smith, Jr., R. Gorlin, M. V. Herman, W. J. Taylor, and J. J. Collins, Jr.: Myocardial Blood Flow in Man: Effects of Coronary Collateral Circulation and Coronary Artery By-Pass Surgery, *J. Clin. Invest.* **51**: 2556, 1972.
80. F. J. Klocke, D. R. Rosing, and D. E. Pittman: Inert Gas Measurements of Coronary Blood Flow, *Am. J. Cardiol.* **23**: 548, 1969.
81. F. J. Klocke, S. M. Whittenberg: Heterogeneity of Coronary Blood Flow in Human Coronary Artery Disease and Experimental Myocardial Infarction, *Am. J. Cardiol.* **24**: 782, 1969.

References

82. C. K. Friedberg, and H. Horn: Acute Myocardial Infarction Not Due to Coronary Artery Occlusion, *J. Am. Assoc.* **112:** 1675, 1939.
83. D. M. Griggs, Jr., and Y. Nakamura: Effects of Coronary Constriction on Myocardial Distribution of Iodoantipyrine-I^{131}, *Am. J. Physiol.* **125:** 1082, 1968.
84. T. W. Moir, and D. W. DeBra: Effect of Left Ventricular Hypertension, Ischemia and Vasoactive Drugs on the Myocardium, *Circulation Res.* **21:** 65. 1967.
85. Cutarelli, R. and Levy, M. N.: Intraventricular Pressure and Distribution of Coronary Blood Flow, *Circulation Res.* **12:** 322, 1963.
86. T. Yipintsoi, W. A. Dobbs, Jr., and J. B. Bassingthwaighte: Regional Deposits of Diffusible Indicators and Microspheres in Dog Left Ventricle. *Federation Procs.* (Abstr.) **31:** 366, 1972.
87. L. C. Becker, N. J. Fortuin, and B. Pitt: Effect of Ischemia and Antianginal Drugs on Distribution of Microspheres in the Canine Left Ventricle, *Circulation Res.* **27:** 263, 1971.

PRINCIPAL EDITOR: FRANK S. CASTELLANA

CONTRIBUTING EDITOR: JAMES B. BASSINGTHWAIGHTE

7. Mathematical Modeling of the Central Circulation

With the continued development of indicator dilution as a diagnostic method for study of the central circulation, techniques of mathematical modeling have become an integral part of procedures used to explain and analyze experimental data. The use of modeling is especially relevant to the interpretation of data from quantitative radiocardiography. The complexity and scope of these data, which include both medium- and high-frequency simultaneous activity–time measurements from several heart chambers, almost precludes the use of any less detailed method of analysis.

Models of the central circulation are used to calculate the system response to a known tracer input or stimulus and may be classified as being either single or multiple component, depending on whether or not discrete segments of the system are lumped together or represented separately. The development and basis of the model equations may be either quasi-empirical or closely related to system anatomy and physiology. In the first case, descriptive modeling, the mathematical representation has as its prime objective an empirical description of system response. As a result, model parameters are not necessarily in one-to-one correspondance with physiological parameters. In the second case, deterministic modeling, the mathematical representation is closely based on system anatomy and physiology, and model parameters have precise, albeit presumed, physiological and anatomical definition. In summary, multiple-component models attempt to describe a physically segmented system as a combination of single-component representations which may be either deterministic or descriptive.

The decision to use a single- or multiple-component approach is affected by the nature of the experimental data and the number of sampling sites available. The first is useful in traditional indicator-dilution studies where a single response to a tracer input function is to be analyzed. Here, a typical experimental situation would be an injection proximal to the right atrium, with sampling from a single site either distal to the left ventricle if a colored dye is used, or over the mid-precordium if a radioactive tracer is employed. In the case of quantitative radiocardiography, where detailed data from discrete cardiopulmonary segments are available, the more complex multiple-component model is preferable if all the data are to be taken into account and the system precisely defined.

The single most important use of modeling is in the estimation of parameter values from experimental data. Parameter estimation entails an iterative procedure of parameter adjustment until the calculated response matches the known system response to a desired degree of accuracy (see Fig. 136). The mathematical techniques of effecting such an analysis are well known (1, 2), and usually involve a least-squares criterion. Thus, given a model

$$\overline{Y}_i = f(X_{i1}, X_{i2}, \ldots X_{im}, b_1, b_2, \ldots b_k)$$

where $\overline{Y}_i$ is the value of the dependent variable (system response) Y, and where there are m independent variables X_{ii}, k parameters b_j, and n observations, the procedure is to adjust the b_j so as to minimize

$$\Phi = \sum_{i=1}^{n} (Y_i - \overline{Y}_i)^2$$

If the residuals are random, it is assumed that the model accurately describes the system and that its parameters can be related either directly or indirectly to physiological or pathological findings. In such analysis the problem of uniqueness must be considered. With a complex model involving a large number of parameters, there is always a possibility that more than one set of parameter values will fit the ex-

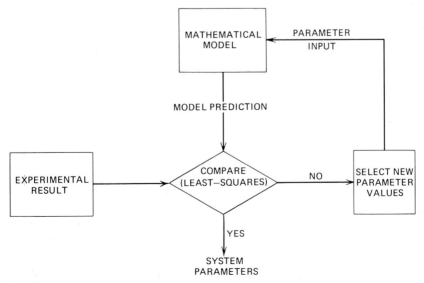

Fig. 136.

perimental data to the desired degree of accuracy. This problem can be overcome in part by carrying out detailed parameter studies and by finding additional constraints on the system from other available data.

The value of a mathematical model does not rest only in its direct application to analysis of experimental data. Rather, a wide range of studies can be executed with the model alone. Such investigations can provide valuable insight into the optimum design of experiments and analysis of data. Three areas are especially relevant: (1) parameter sensitivity studies, (2) generation of simulated experimental data, and (3) simulated testing of new experimental techniques.

Parameter sensitivity studies are used to determine the effect of a change in one parameter on the overall system response to a specific input. They provide the investigator with a knowledge of which parameters most critically affect his system. Such studies may suggest simplifications that result in both ease of computation and more precise definition of important variables. Models can also be employed to generate simulated experimental data that can be used to verify techniques of analysis and to describe system response at known conditions and in the presence of abnormalities. Such results provide the investigator with prior knowledge of how the system should respond under certain conditions and to what extent changes in these conditions or the presence of abnormalities will be observable. One might, for example, model an atrial septal defect in order to generate simulated experimental data either to test a proposed theoretical analysis or to observe the result of the defect on precordial dilution curves from one or more heart chambers. Visualization of the latter would be especially valuable in determining optimal measuring locations for *in vivo* studies. Finally, models can be used to partially test new experimental techniques by providing verification prior to any *in vivo* or patient studies. One might easily study the relative merits of either a peripheral vein or direct right atrial injection, or one might assess the value of a double pulse injection technique for the direct measurement of transit times.

The section that follows presents a brief overview of models in both classes with no attempt at a complete survey. Rather, the objective is to build a format for the detailed development of a general multiple-component compartmental model that should provide the reader with further insight into the form and the analysis of experimental data that might be expected from a high-frequency study with a gamma-camera in both the normal and the abnormal patient. It is recognized that the model is by no means rigorous and contains several implicit assumptions demanding further justification; these will be discussed. The last section considers problems related to the analysis of experimental data.

SINGLE- AND MULTIPLE-COMPONENT MODELS OF THE CENTRAL CIRCULATION

Multiple-parameter single-component models of the central circulation are designed to interpret experimental data from a single sampling site that result in response to a defined tracer input (Figure 137). This class of model finds application in the analysis of data from traditional dye-dilution studies, but is less suited to analysis of more detailed data from multiple sampling sites that is now available from quantitative radiocardiography. The component may describe the entire central cir-

Single- and Multiple-Component Models of the Central Circulation

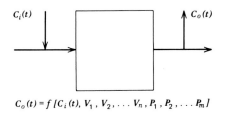

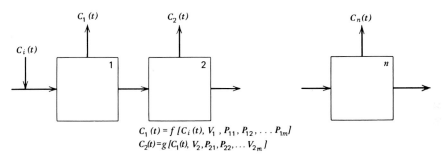

Fig. 137. Functional representation of single- and multiple-component systems.

culation or it may describe a segment of the central circulation such as the right heart or lungs. The structure of the individual component can assume various forms. Harris and Newman (3), in a recent review article, categorize two of these as (*1*) compartmental and (*2*) dispersion. The development of both is related to fundamental mechanisms that affect the time distribution of tracer in the cardio-pulmonary circulation. A third form unrelated to mechanisms employs empirical curve-fitting techniques using specific functions (random walk, etc.) to fit the experimental data.

Compartmental models are formulated with the assumption that relatively good mixing occurs in certain segments of the circulation (atria and ventricles) and that in other segments (arteries, veins, and lung capillary beds) mixing is relatively poor. One or more perfect mixers and/or time delays are used in series or series–parallel combination to simulate specific segments of interest. A mathematical representation is effected by writing a tracer mass balance equation for each compartment of the model. The resulting set of coupled simultaneous linear first-order differential equations is solved for the output response to a specified input. The continuous flow, constant volume mass balance equations for a perfect mixer and a pure delay are, respectively,

$$QC_{i-1} - QC_i = V_i \left(\frac{dC_i}{dt}\right) \tag{1}$$

$$C_i(t) = C_{i-1}(t - t_d) = C_{i-1}\left(t - \frac{V_i}{Q}\right) \tag{2}$$

where C_i is the concentration of tracer in the ith mixer (m/l^3), Q the volumetric flow rate (l^3/t), V_i the volume of the ith mixer (l^3), and t_d the transport delay (t). The number of compartments can vary from one to several (4–6) and the system equations can be solved in the time domain (5, 7) or in the frequency domain using Laplace (8) or Fourier (9) transforms.

Two examples of models that may be related to this category are the series mixer model of Schlossmacher et al. (60) and Bassingthwaighte's lagged normal density model (4). Schlossmacher considers n equal-sized perfect mixers in series and develops an equation for the system response to an ideal impulse. The equation describes a family of curves that vary from an exponential decay for $n = 1$, to a unit impulse function for $n = \infty$. Bassingthwaighte et al. (4) consider the equivalent of a single perfect mixer and an input composed of a Gaussian distribution function. The model response is a "lagged normal density curve" determined by four parameters. While these parameters do not identify with a one-to-one correspondence a specific chamber or distribution of path lengths, the authors show that they can be related to physiological values of interest for a number of different situations. Figure 138 illustrates the ranges of response of these models.

Dispersion models treat mixing as uniformly distributed throughout the system rather than at discrete locations as in the compartmental approach. When an indicator is injected into a moving stream, dispersion occurs as the result of radial and longitudinal diffusion and longitudinal convection. For a single tracer (a) injected into a flowing stream (b), dispersion is characterized by the equation of continuity for a binary mixture (10):

$$\frac{\partial C_a}{\partial t} + (\mathbf{v} \cdot \nabla C_a) = D_{ab}\nabla^2 C_a + R_a \tag{4}$$

C_a is the tracer concentration (m/l^3), $\mathbf{v}$ the stream velocity vector (l/t), D_{ab} the mass diffusivity (l^2/t), and R_a the rate of tracer production [m/(l^{3}t)]. For $R_a = 0$ and simple one-dimensional transport, Equation 4 reduces to:

$$\frac{\partial C_a}{\partial t} + v_x \left(\frac{\partial C_a}{\partial x}\right) = D_{ab}\left(\frac{\partial^2 C_a}{\partial x^2}\right) \tag{5}$$

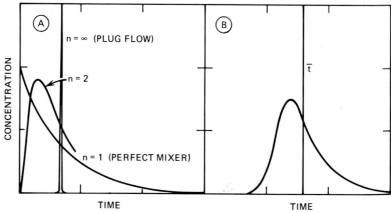

Fig. 138. Typical output response of two-compartment models: (*A*) n perfect mixers (Ref. 6); (*B*) lagged normal density curve (Ref. 4).

Single- and Multiple-Component Models of the Central Circulation

For more complex systems Equation 4 is still used, but with the mass diffusivity D_{ab} replaced by an "equivalent dispersivity," D, determined by geometry and specific flow characteristics. The application of this equation to physical systems has been considered by several investigators (11, 12). In the analysis of dilution curves v_x is presumed to be related to the cardiac output and D to the anatomical structure of the heart. As D approaches ∞ or 0, the dispersion model approaches a single perfect mixer or a pure time delay (13) (Fig. 139).

Multiple-component models of the central circulation are designed to interpret experimental data from multiple sampling sites (Fig. 137). This class of model is especially applicable to analysis of dynamic data from a gamma-camera or related device, since these instruments are capable of measuring regional simultaneous activity–time functions of an injected gamma-emitting tracer and also provide data at sufficiently high count rates to define intrabeat events of the cardiac cycle.

The most popular approach to individual component modeling in the multiple-component case has been the compartmental approach. The multiple-compartment models thus formed differ from those previously considered in that each compartment has a presumed one-to-one correspondance with an individual heart chamber or combination of chambers. Thus, the measured activity from a discrete region of the heart is directly related to the calculated tracer concentration in either one or a combination of several model compartments. As a first approximation, this approach appears well suited to modeling the discrete chamber geometry of the heart. The complex parallel capillary structure of the lungs, however, precludes a simple one-to-one modeling in terms of perfect mixers. Since precise regional data

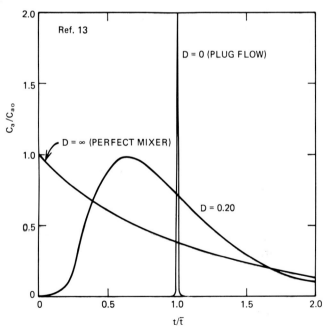

Fig. 139. Typical output response of a dispersion model: $D = 0$, 0.020 and ∞ (Ref. 13).

are not available in any case, an adequate alternative would be a simulation of the distribution of lung transit times in terms of another form of single-component model (e.g., see ref. 20).

The first models to appear were continuous-flow, constant volume approximations. Newman et al. (5) simulated the right heart, lungs, and left heart as three compartments in series. This model did not identify individual heart chambers or account adequately for transport delay in the complex lung capillary system. A more realistic approximation was suggested by Shames (17), who introduced two additional compartments to represent the atria and ventricles separately. Ishii and MacIntyre (18) recently employed an analog computer to simulate a continuous-flow approximation consisting of a four-compartment heart, and a lung represented by a series combination of a time delay and a single mixed compartment.

DEVELOPMENT OF A GENERAL PULSATILE MODEL OF THE CENTRAL CIRCULATION

While the continuous-flow multiple component models may be adequate for the description of medium frequency time smoothed data, they are incapable of simulating the fine structure of the cardiac cycle, and lead to several anomalies such as the coincidental appearance of an injected tracer in all catenary compartments not isolated by time delays. Since data concerning the high frequency events are now available, and in all likelihood have the potential to provide a valuable measure of cardiac performance, it is apparent that a more general pulsatile model that can simulate discontinuous flow as well as account for volume variations during systole and diastole is desirable.

The theory of compartmental modeling is predicated on the assumption of perfect mixing. While considerable attention has been focused on this assumption, no definitive conclusions have as yet been drawn. Many investigators report poor left ventricular mixing in studies with a radiopaque dye and cine-angiography. In this situation a relatively large quantity of highly viscous material is injected, and effective mixing is not likely to occur as the magnitude of turbulent transport is severely reduced. In light of the relatively small amount of nonviscous isotopic tracer injected in a radiocardiography study, the assumption of good mixing appears more tenable at least as a first approximation. Unpublished *in vitro* mixing experiments at St. Luke's Hospital Center (19) that considered tracer injection into a stagnant pool seem to support this assumption. A possible approach in a more complex development would be to subdivide a chamber into several regions and consider finite rates of turbulent inter-region transport. This would permit a more detailed description of mixing but at the expense of additional system parameters.

Consider the compartmental model representation of the cardiopulmonary circulation shown in Figure 140*a*. The heart is simulated as four separate compartments, and the lungs as a single compartment in series with a time delay. Compartments *1* and *5*, the right and left atria, and compartment *3*, the lung time delay, are considered to be of constant volume. Compartments *2* and *6*, the right and left ventricles, have volumes which vary in a prescribed fashion during diastole and systole. The compartmental relation to the four-chamber heart anatomy is obvious, and since the actual atrial volume variation is relatively small, the initial

Development of a General Pulsatile Model of the Central Circulation

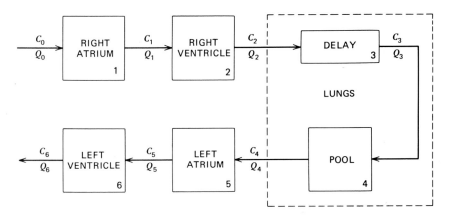

Fig. 140a. A model for the central circulation.

approximation of constant volume appears reasonable. Because of the lung's complex anatomic definition, its representation is a more complex problem. The use of a single compartment is inadequate because of the time delay introduced by the presence of the capillary bed. The decision to employ a time delay in series with a mixed chamber of varying volume was based on the assumption that this closely resembles the *in vivo* situation, where one could expect turbulence and good mixing in the large arteries and veins entering and exiting the lungs, but a virtual time delay in the complex parallel-flow capillary system. Lung compliance is accounted for in the varying volume chamber.

During diastole, the ventricular exit flows Q_2 and Q_6 as well as the lung flow Q_3 are equal to 0, and the inlet flows Q_1 and Q_5 are equal to the instantaneous time rate of change of ventricular volume, dV_i/dt. Since the atrial volumes are constant, their inlet and exit flows are identical and equal to the instantaneous time rate of change of volume of their respective ventricles. During systole the ventricular inlet flows Q_1 and Q_5 are equal to 0, and the ventricular exit flows Q_2 and Q_6

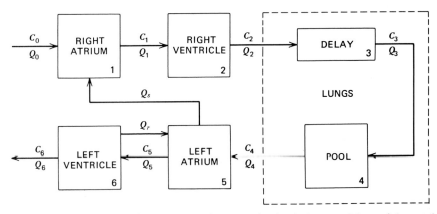

Fig. 140b. Shunts superimposed on the central circulation model: atrial septal defect and mitral insufficiency are considered.

are equal to the instantaneous time rate of change of ventricular volume during systole; flow into lung compartment 4 is equal to Q_2, and both atrial flows are equal to 0. The volume and flow functions for each chamber during diastole and systole are summarized in Table 6. The flow functions $(dV_i/dt)_d$ and $(dV_i/dt)_s$ in this table are determined from the instantaneous time varying volumes of the right and left ventricles during diastole and systole.

Table 6. Volume and Flow Functions During Diastole and Systole.[a]

Chamber	Diastole ($T_{es} < t \leq T_{ed}$)		Systole ($T_{ed} < t \leq T_{es}$)	
	V_i	Q_i	V_i	Q_i
1	V_1	$(dV_2/dt)_d$	V_1	0
2	$V_{2es} + V_{2d}(t)$	0	$V_{2ed} - V_{2s}(t)$	$(dV_2/dt)_s$
3	V_3	0	V_3	$(dV_2/dt)_s$
4	$V_{4ed} - V_{4d}(t)$	$(dV_4/dt)_d$	$V_{4es} + V_{4s}(t)$	0
5	V_5	$(dV_6/dt)_d$	V_5	0
6	$V_{6es} + V_{6d}(t)$	0	$V_{6ed} - V_{6s}(t)$	$(dV_6/dt)_s$

[a] V_{ies} = volume of Chamber i at end systole; V_{ied} = volume of Chamber i at end diastole; $V_{is}(t)$ = volume variation function of Chamber i during systole; $V_{id}(t)$ = volume variation function of Chamber i during diastole; Q_i = rate of exit flow from Chamber i.

The mass balance equations for each compartment are given by Equations 6–11.

$$Q_0 C_0 - Q_1 C_1 = \frac{d(C_1 V_1)}{dt} = V_1 \left(\frac{dC_1}{dt}\right) \tag{6}$$

$$Q_1 C_1 - Q_2 C_2 = \frac{d(C_2 V_2)}{dt} = C_2 \left(\frac{dV_2}{dt}\right) + V_2 \left(\frac{dC_2}{dt}\right) \tag{7}$$

$$C_3(t) = C_2 \left(t - \frac{V_3}{Q_3}\right) \tag{8}$$

$$Q_3 C_3 - Q_4 C_4 = \frac{d(C_4 V_4)}{dt} = C_4 \left(\frac{dV_4}{dt}\right) + V_4 \left(\frac{dC_4}{dt}\right) \tag{9}$$

$$Q_4 C_4 - Q_5 C_5 = \frac{d(C_5 V_5)}{dt} = V_5 \left(\frac{dC_5}{dt}\right) \tag{10}$$

$$Q_5 C_5 - Q_6 C_6 = \frac{d(C_6 V_6)}{dt} = C_6 \left(\frac{dV_6}{dt}\right) + V_6 \left(\frac{dC_6}{dt}\right) \tag{11}$$

In these equations C_i is the tracer concentration, V_i the compartment volume, and Q_i the exit flow rate. As the system is defined

$$Q_0(t) = Q_1(t)$$
$$Q_2(t) = Q_3(t)$$
$$Q_4(t) = Q_5(t)$$

and

$$\int_{T_{es}}^{T_{ed}} Q_i dt = \int_{T_{ed}}^{T_{es}} Q_i dt \qquad i = 2,4,6$$

The injected tracer is represented as an additional input function, $C_I Q_I$, to one of the compartments. During diastole the ventricular exit flow, Q_i, is 0, and the inlet flow is equal to the time rate of change of ventricular volume, dV_i/dt. The general mass balance equation for a compartment of varying volume then becomes

$$\left(\frac{dV_i}{dt}\right)(C_i - C_{i-1}) + V_i\left(\frac{dC_i}{dt}\right) = 0 \qquad (12)$$

During systole the inlet flow is 0 and the exit flow is equal to minus the time rate of change of ventricular volume. The general mass balance equation then reduces to

$$\frac{dC_i}{dt} = 0 \qquad (13)$$

and the concentration is constant.

To solve Equations 6–11, the volume variations of compartments 2, 4, and 6 with time during diastole and systole must be known. As a first approximation, a linear variation may be considered:

$$V_i = V_{ies} + at \qquad T_{es} \le t \le T_{ed} \qquad (14)$$

$$V_i = V_{ied} - b(t - T_{ed}) \qquad T_{ed} < t < T_{es} \qquad (15)$$

where V_{ies} and V_{ied} are, respectively, end systolic and end diastolic compartment volumes. An improved approximation can be obtained by introducing an additional parameter and considering, for example, an exponential filling phase and a linear emptying phase

$$V_i = V_{ies} + a(1 - e^{-ct}) \qquad T_{es} \le t \le T_{ed} \qquad (16)$$

$$V_i = V_{ied} - b(t - T_{ed}) \qquad T_{ed} < t < T_{es} \qquad (17)$$

As more data become available it may be convenient to represent any arbitrary cycle as a polynomial function of degree n

$$V_i = a + bt + ct^2 + \ldots + nt^{(n-1)} \qquad (18)$$

Once the form of the volume–time function is selected for each ventricle, Equations 6–11 are integrated forward in time over one cycle. Selection of the appropriate volume function during the integration depends on whether the phase of the cycle is systolic or diastolic. The integration is affected numerically using finite differences. Initial conditions for the first cycle are determined from the form and location of the tracer input function. In subsequent cycles, the compartment concentrations at end systole for cycle n are used as initial conditions for cycle $n+1$. Forward integration during each cycle follows the constraints detailed in Table 6.

The mathematical model as developed serves two functions: (*1*) given cardiac output, compartment volumes, and the required volume–time functions, the model will generate the expected response to a specific tracer input function; and (*2*) given experimental data, the model can be used in conjunction with an appropriate regression analysis to determine best-fit values of the model parameters.

Model Representation of Cardiopulmonary Abnormalities

The effects of various cardiopulmonary abnormalities on the regional and overall activity–time functions are determined by appropriate modification of the model shown in Figure 140a described by Equations 6–11. To illustrate the method, two specific abnormalities are considered: (1) an atrial septal defect and (2) an insufficient mitral valve. The effect of each on the model is shown diagramatically in Figure 140b.

The Atrial Septal Defect

An atrial–septal defect refers to a relatively large nonvalvular opening in the atrial septum through which oxygenated blood from the left heart is added to the normal intake of the right atrium. The flow, usually as much as two to three times cardiac output, occurs mainly during the period of ventricular filling immediately after the opening of the tricuspid valve (21). The shunt results in progressive right ventricular enlargement, decreased left ventricular stroke output, and eventual failure. In modeling an atrial septal defect the following assumptions are made: (1) the direction of flow is always from the left to right atrium, (2) flow occurs only during the diastolic filling period, and at a constant rate, and (3) atrial volumes remain constant, the additional shunt flow being accommodated by an increased right ventricular end diastolic volume. The presence of the shunt requires modification of the tracer mass balance equations for compartments 1 and 5 (see Fig. 140b). These equations become, respectively.

$$Q_0 C_0 + Q_s C_5 - Q_1 C_1 = \frac{d(C_1 V_1)}{dt} = V_1 \left(\frac{dC_1}{dt} \right) \quad (6')$$

$$Q_4 C_4 - Q_5 C_5 - Q_s C_5 = \frac{d(C_5 V_5)}{dt} = V_5 \left(\frac{dC_5}{dt} \right) \quad (10')$$

where Q_s is the shunt flow from left to right atrium. In accord with assumption 2

$$Q_s = 0 \qquad T_{ed} \leq t \leq T_{es}$$

$$Q_s = \frac{dV_2}{dt} - Q_0 = \text{constant} \qquad T_{es} < t < T_{ed}$$

The above modified variations of Equations 6–11 are integrated numerically as previously described.

Mitral Insufficiency

Mitral insufficiency refers to a closing defect of the mitral valve. During left ventricular systole, a fraction of forward flow leaks back into the left atrium, thereby increasing its volume and the pressure within it (21). During diastole, the left ventricle is subject to the relatively high filling pressure of the atrium and therefore dilates to accommodate the regurgitant flow that had leaked into the atrium during the previous cycle. The stroke volume of the left ventricle is increased by an amount equal to the regurgitant flow, while cardiac output remains constant. In modeling mitral incompetence the following assumptions are made: (*1*) regurgitation occurs during ventricular systole at a constant rate, (*2*) the regurgitant flow accumulated during systole is all accommodated by the left atrium, and the pulmonary system remains unaffected, and (*3*) during diastole the rate of regurgitant flow into the ventricle is constant, the accumulated flow being accommodated by an appropriately increased left ventricular end diastolic volume. The presence of the insufficient valve requires modification of the tracer mass balance equations for compartments 5 and 6 (see Fig. 140*b*). These equations become, respectively,

$$Q_4 C_4 + Q_r C_6 - Q_5 C_5 = V_5 \left(\frac{dC_5}{dt}\right) + C_5 \left(\frac{dV_5}{dt}\right) \quad (10'')$$

$$Q_5 C_5 - Q_r C_6 - Q_6 C_6 = V_6 \left(\frac{dC_6}{dt}\right) + C_6 \left(\frac{dV_6}{dt}\right) \quad (11'')$$

where Q_r is the regurgitant flow from left ventricle to atrium. In accord with the assumptions

$$Q_r = 0 \qquad\qquad T_{es} \leq t \leq T_{ed}$$

$$Q_r = -\frac{dV_6}{dt} - Q_6 = \text{constant} \qquad T_{ed} < t < T_{es}$$

As in the previous case, the modified mass balance equations are integrated numerically.

Solution of the Model Equations

Using the values of the cardiopulmonary parameters detailed in Table 7, the system response to right atrial tracer injection during ventricular systole was evaluated. The ventricular volume function selected was that described by Equations 14 and 15. The resulting flow and volume variations over one cardiac cycle are

Table 7. Test Parameters

Cardiac output (Q)	4.8 liter/minute
Stroke volume	80 ml
Right atrial volume (V_1)	110 ml
Right ventricular end systolic volume (V_{2es})	80 ml
Lung time delay (V_3/Q)	3 seconds
Lung volume—compartment 4 (V_{4es})	160 ml
Total lung volume ($V_3 + V_{4es}$)	400 ml
Left atrial volume (V_5)	110 ml
Left ventricular end systolic volume (V_{6es})	45 ml
Injection duration	0.20 second
Mass injected (I)	0.20 g
Counting efficiency constant	100,000 cps/g

shown in Figure 141. It should be noted that compartment concentrations at end systole and end diastole are independent of the form of the volume function selected. For a heart rate of 60 beats/minute, the diastolic filling period was set at 0.70 seconds and the systolic emptying period at 0.30 seconds. The stroke volume of 80 ml represented a left ventricular ejection fraction of 0.64. The tracer input function was a unit step of 0.20 seconds duration directly into the right atrium at the start of ventricular systole ($0.70 \leq t \leq 0.90$).

Equations 6–11 were integrated over 25 cardiac cycles employing a fourth-order Runge-Kutta integration procedure with an error of order h^5 (step size) and a step size of 0.05 seconds. Integration of the equations was effected separately for systole and diastole, and a comparison of step sizes indicated the adequacy of an h of 0.05 seconds.

Data from the gamma-camera and related sampling instruments are usually presented as the number of radioactive counts, a_i, in a specified period called the frame time.

$$a_i = \int_0^{T_f} (\alpha_i C_i V_i)\, dt$$

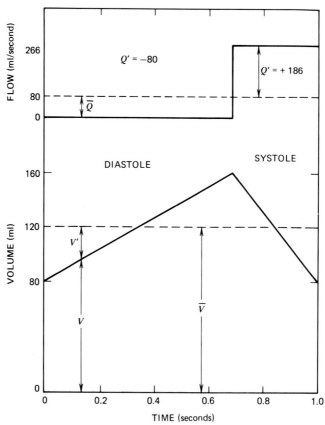

Fig. 141. Time–volume functions in the pulsatile model system, indicating flow rate and volume.

Here, a_i is the activity in counts per frame time, T_f is the frame time in seconds, and α_i is the counting efficiency constant in counts per second per gram. Using the indicated value of $\alpha = 100{,}000$ cps/g, the results of the integration for frame times of 0.10, 0.20, and 0.50 seconds were determined and are shown plotted in Figure 142. It should be noted that in these and subsequent figures the plotted frame activities were connected by straight lines for ease of visualization.

The activity fluctuations that result from systole and diastole, as well as the apparent smoothing effect of an increased frame time are manifest in the Figures shown. It is seen that even with a simple linear filling and emptying function, a qualitative comparison with available experimental data (23) is quite favorable. Note particularly that in the third panel the shape of the total heart curve is similar to that expected from standard single-probe analysis.

At higher or lower flow rates the shapes of the output curves are similar, that is they are only compressed or expanded in time. This results from the assumption of perfect mixing and the presumed independence of mixing and flow rate. If mixing were not perfect, a variation of the output response with flow would certainly be expected and similarity would not occur. If similarity does in fact exist, it has the greatest likelihood of being found in a system limited to the major cardiac chambers rather than in one encompassing a large volume of peripheral vessels. In such vessels the moderating influence of the wall is significant and the resulting effect of flow on turbulent transport more pronounced.

Predicted Response in the Presence of Cardiac Abnormalities

Figure 142 shows the predicted response of a "normal" heart to right atrial tracer injection. The value of both the method and the model, however, is in the elucidation and quantification of cardiac abnormalities. Patient cardiac response in the presence of such abnormalities has already been documented by several investigators (24, 25). Using the model, the predicted responses in the presence of various diseases of the heart can be categorized both qualitatively and quantitatively, and used as both teaching and diagnostic aids. Three abnormalities are considered: (*1*) left ventricular failure, (*2*) atrial septal defect, and (*3*) mitral insufficiency.

Left Ventricular Failure

In a condition approaching pure left ventricular failure, cardiac output is usually maintained at the expense of a dilated left ventricle. As a consequence, left ventricular ejection fraction is markedly reduced and the left heart mean transit time increased; an increased pulmonary blood volume is also usually present. Figure 143 shows the model response in the form of activity–time curves for the left ventricle, the left ventricle plus atrium, and the combined left and right hearts, in the presence of a dilated left ventricle having an end diastolic volume of 240 ml. When this result is compared with the "normal" response of Figure 142, the expected decrease in ejection fraction and increase in transit time are manifest as significantly smaller fluctuations in the left ventricular activity–time curve and an increase and prolonging of the level of activity. These results agree with clinical findings.

Atrial Septal Defect

A description of the atrial septal defect and its mathematical representation has already been presented. Figures 144 and 145 show the expected right and left heart responses in the presence of a shunt flow equal to cardiac output. The result of the defect is manifest in both figures as an increased right ventricular ejection

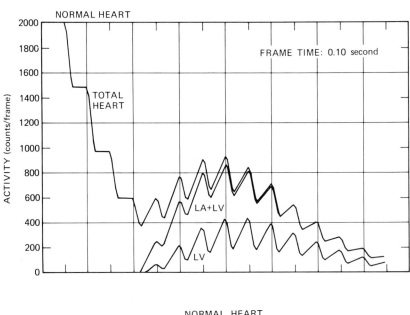

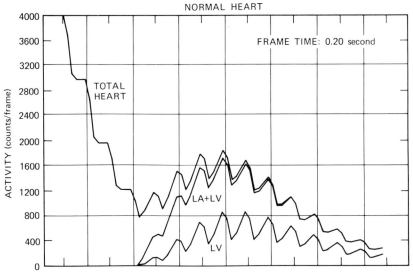

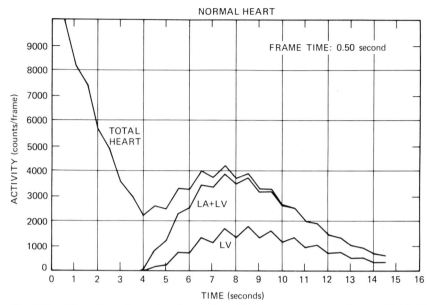

Fig. 142. The effect of varying frame time. Test parameters are given in Table 7. Frame times are 0.1, 0.2, and 0.5 seconds.

fraction, an increased overall transit time resulting from shunt recirculation, and a rapid recirculation peak in the right atrial activity–time curve.

Mitral Insufficiency

Figure 146 shows the expected total- and left-heart responses in the presence of an insufficient mitral valve that permits a regurgitant fraction equal to cardiac output. As expected, this defect shows up as an increased left ventricular ejection fraction and an increased left heart mean transit time.

The Continuous-Flow Approximation

Equations 6–11 were developed for the general case of a varying-volume pulsatile-flow situation. The continuous-flow constant-volume approximation represents the special case where

$$\frac{dV_2}{dt} = \frac{dV_4}{dt} = \frac{dV_6}{dt} = 0$$

$$Q_0(t) = Q_1(t) = \ldots = Q_6(t) = \text{constant}$$

The equations were solved for these conditions and the parameter values of Table 7. The result is shown in Figure 147. Activity fluctuations resulting from systole and diastole are not predicted, and it is observed, as expected, that tracer appears simultaneously in all compartments not separated by time delays.

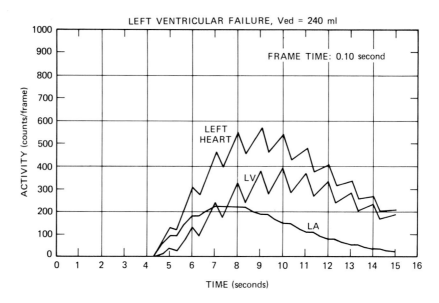

Fig. 143. Model response with an increase in left ventricular volume. Time frame = 0.05 second, V_{es} = 200 ml, V_{ed} = 280 ml, and $\bar{V}$ = 240 ml.

ANALYSIS OF EXPERIMENTAL DATA

Data from gamma-camera radiocardiography are generally available in two basic forms: (*1*) activity data from a subregion of a specific compartment and (*2*) total activity data from an entire compartment. The first are proportional to tracer *concentration* if the volume considered does not vary with time, and the second are proportional to *tracer mass*, or the product of tracer concentration and compartment volume. The efficacy and precision with which these data are obtained is a complex function of camera and tracer characteristics, and the relative positions of the heart and the camera when the data are taken. Because of the relative proximity of the heart chambers and the resultant effects of scattering, and because of partial chamber overlap in certain fields of view, the measured and actual chamber activity may differ considerably. This effect, commonly termed cross-talk, must be taken into account if the data are to be interpreted correctly (17).

The scintillation camera measures isotope activity in a volume region of interest. The relation between measured activity and tracer concentration is described by Equation 19

$$a_i = \alpha_i V_i C_i \qquad (19)$$

where a_i is the measured activity, V_i the volume observed by the detector, C_i the tracer concentration in that volume, and α_i a counting efficiency constant. The volume observed by the detector may be an entire chamber or part of a chamber or several overlapping chambers and may be constant or may vary during systole and

Analysis of Experimental Data

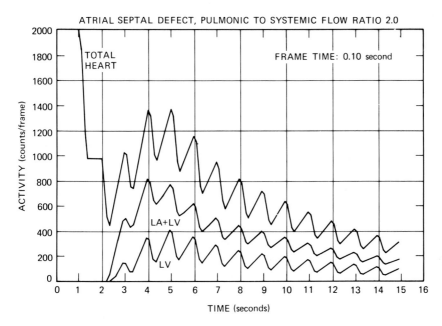

Fig. 144. Atrial septal defect, shunt flow equals cardiac output: left-heart response.

diastole. As a first approximation the atria, pulmonary artery, and aorta are usually considered to have constant volume. Since both C_i and V_i are functions of time

$$\frac{da_i}{dt} = \alpha_i V_i \left(\frac{dC_i}{dt}\right) + \alpha_i C_i \left(\frac{dV_i}{dt}\right) \qquad (20)$$

and activity variations are related to both concentration and volume changes. If the observed volume is constant

$$\frac{da_i}{dt} = \alpha_i V_i \left(\frac{dC_i}{dt}\right) \qquad (21)$$

and changes in activity are directly related to changes in concentration. During systole, ventricular concentration is constant and changes in activity are directly proportional to volume variations

$$\frac{da_i}{dt} = \alpha_i C_i \left(\frac{dV_i}{dt}\right) \qquad (22)$$

Possible regions of interest for a single ventricular chamber are illustrated in Figure 148. The broken line represents a planar projection of the chamber boundary at end diastole, and the solid line a planar projection at end systole. If region of interest *1* were selected, the measured activity would vary from some positive value at end diastole to 0 at end systole, and the calculated ejection fraction would be 1. Region of interest *2* is totally within the bounds of the projected end systolic volume, and any activity variations during systole are related (see Equation 22) to ventricular volume variations in a direction normal to the camera crystal. An

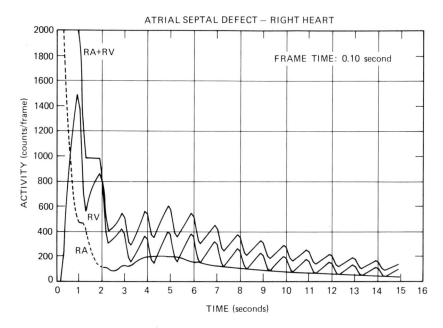

Fig. 145. Identical lesion: right-heart response. Note early circulation.

ejection fraction calculated using data from this region would only equal the true ventricular ejection fraction if the region of interest sampled were representative of the entire ventricle, and this in general is unlikely. Region of interest *3* encompasses the entire ventricle during both systole and diastole. Activity fluctuations are described by Equation 20, and, specifically, during systole by Equation 22 from which the true ejection fraction is calculated. As this example illustrates, the importance of a correct region of interest selection cannot be overemphasized.

In the development thus far only a single chamber has been considered. In practice, however, the camera may be measuring activity from a composite of the region of interest and overlapping regions, as well as scattered radiation from adjacent regions. This situation is described mathematically by modifying Equation 19 to include two additional terms:

$$a_i = \alpha_i V_i C_i + \sum \alpha_j V_j C_j + \sum \beta_{ji} m_j \qquad (23)$$

C_j is the tracer concentration in overlapping chamber j, V_j the volume of chamber j observed by the detector, and m_j the mass of tracer in an adjacent chamber j from which scattering occurs, and α_j and β_{ji} are counting efficiency constants. The contributions are described diagramatically in Figure 149. The analysis of experimental data is far more complex for the general situation described by Equation 23. Activity variations in any region of interest are attributable to multiple sources, and it becomes difficult, if not impossible, to relate these to volume variations from a specific chamber without the aid of additional measurements and a conceptual mathematical model to resolve all the data. This problem is addressed by Shames (17) in his development of a five-compartment continuous flow model.

At equilibrium

$$\bar{a}_i = \alpha_i V_i \bar{C} + \sum \alpha_j V_j \bar{C} + \sum \beta_{ji} \bar{m}_j \qquad (24)$$

Analysis of Experimental Data

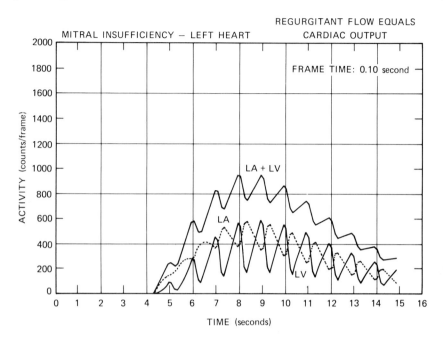

Fig. 146. Mitral insufficiency, regurgitant flow equals cardiac output.

where $\overline{C}$ is the equilibrium tracer concentration, $\bar{a}_i$ the equilibrium activity measured from chamber i, and $\overline{m}_j$ the mass of tracer in compartment j at equilibrium. If regions can be found such that $V_j \doteq 0$ and $\beta_{ji} = 0$ (no overlapping chambers and no scattering), it then follows that

$$C_i = a_i \left(\frac{\overline{C}}{\bar{a}_i}\right) \quad (25)$$

Using this relation, with a knowledge of $\overline{C}$ and $\bar{a}_i$, Equations 6–11 can be solved directly in terms of measured activities. It should be noted, however, that for varying volume chambers $\bar{a}_i$ is a periodic function of time and should be included as such in Equation 25.

Almost all current investigations are concerned with the analysis of time-smoothed data. The chamber volumes calculated from these data, however, are not clearly defined. They may represent average, end systolic or end diastolic volumes, or something else entirely. To interpret these time-smoothed quantities we replace activity, concentration, and volume by an average plus a fluctuating value. This is a usual procedure and finds a direct parallel in the theory of turbulence. Thus

$$a_i = \bar{a}_i + a_i' \quad (26)$$

$$C_i = \overline{C}_i + C_i' \quad (27)$$

$$V_i = \overline{V}_i + V_i' \quad (28)$$

where a_i, C_i, and V_i are instantaneous values, $\bar{a}_i$, $\overline{C}_i$, and $\overline{V}_i$ are time-smoothed and a_i', C_i', and V_i' are instantaneous fluctuations. It should be noted that the time-averaged values also vary with time, but with a significantly longer period, and

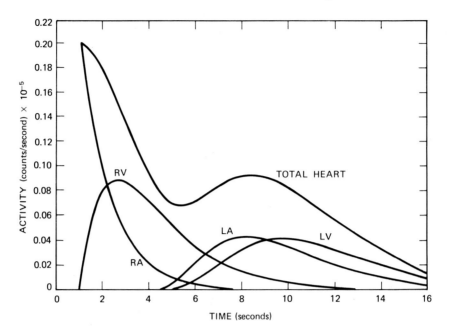

Fig. 147. Response of the model when a nonpulsatile (continuous flow) model is used.

depend for their exact form on the instrument averaging procedures employed. The time averages of the fluctuations by definition are zero.

$$\int a_i' dt = \int C_i' dt = \int V_i' dt = 0 \qquad (29)$$

Substituting for the instantaneous values in Equation 19

$$\bar{a}_i + a_i' = \alpha_i (\bar{V}_i + V_i')(\bar{C}_i + C_i') \qquad (30)$$

expanding

$$a_i + a_i' = \alpha_i \bar{V}_i \bar{C}_i + \alpha_i \bar{V}_i C_i' + \alpha_i V_i' \bar{C}_i + \alpha_i V_i' C_i' \qquad (31)$$

and then time averaging each term, we obtain a relation between time-averaged activity, volume, and concentration.

$$\bar{a}_i = \alpha_i \overline{V_i C_i} + \alpha_i \overline{V_i' C_i'} \qquad (32)$$

It is usual to assume that $\overline{V_i' C_i'}$ is small in comparison to $\overline{V_i C_i}$, and that average activity can be related directly to average concentration by the relation

$$\bar{a}_i = \alpha_i \overline{V_i C_i} \qquad (33)$$

In actuality the magnitude of $\overline{V_i' C_i'}$ may be significant, and the validity of Equation 33 is open to question.

In the analysis of data the problem of recirculation must be considered if the time for reappearance of tracer is on the order of the transit time through the system. If recirculation is a late event, exponential extrapolation of the output curve is a reasonable procedure (14). If, however, recirculation is early enough to

Analysis of Experimental Data

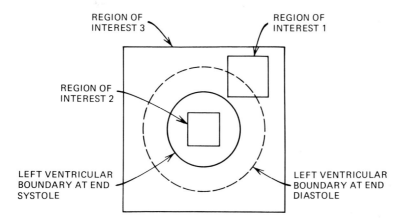

Fig. 148. Stylized regions of interest which may be selected for different purposes in radiocardiography.

severely distort the primary response curve, other procedures must be considered (15). These include the use of multiple output measurements (16) and a direct accounting for recirculated tracer in the mass balance equations through use of a coupled model for the systemic circulation (7).

NOMENCLATURE

- α Counting efficiency constant
- β Cross-talk constant
- C Tracer concentration (g/ml)
- Q Flow (ml/minute)
- t Time (minutes)
- V Compartment volume (ml)

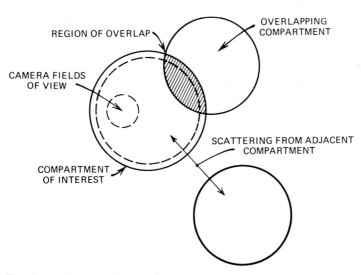

Fig. 149. Sources of counts in a region of interest.

Subscripts

d   Occurring during diastole
es End systole
ed End diastole
I   Referring to injected tracer
i   Referring to compartment i
s   Occurring during systole

REFERENCES

1. D. W. Marquardt, *J. Soc. Ind. Appl. Math.*, **11**: 431, 1963.
2. H. H. Rosenbrock, and C. Storey: *Computational Techniques for Chemical Engineers*, Pergamon Press, Oxford, 1966, Chap. 8.
3. T. R. Harris, and E. V. Newman: An Analysis of Mathematical Models of Circulatory Indicator Dilution Curves, *J. Appl. Physiol.* **28**: 840, 1970.
4. J. B. Bassingthwaighte, F. H. Ackerman, and E. H. Wood: Applications of the Lagged Normal Density Curve as a Model for Arterial Dilution Curves, *Circ. Res.* **18**: 398, 1966.
5. E. V. Newman, et al.: The Dye Dilution Method for Describing the Central Circulation: An Analysis of Factors Shaping the Time-Concentration Curve, *Circulation* 4: 735, 1971.
6. E. J. Schlossmacher, et al.: Perfect Mixers in Series Model for Fitting Venoarterial Indicator-Dilution Curves, *J. Appl. Physiol.* **22**: 327, 1967.
7. R. R. Lunt, Cardiac Output Estimation from Indicator-Dilution Data, MS Thesis, Columbia University, 1969.
8. C. W. Sheppard, and G. B. Spurr, Periodic and Aperiodic Phases of Circulatory Mixing, *Bull. Math. Biophys.* **27**: 65, 1965.
9. T. R. Harris, The Identification of Recirculating Systems in the Frequency Domain, *Bull. Math. Biophys.* **30**: 87, 1968.
10. R. B. Bird, W. E. Stewart, and E. N. Lightfoot, *Transport Phenomena,* Wiley, New York, 1960.
11. G. I. Taylor, The Dispersion of Matter in Turbulent Flow Through a Pipe, *Proc. Roy. Soc., London, Ser. A* **223**: 446, 1954.
12. O. Levenspiel, and K. B. Bischoff: *Advan. Chem. Eng.* **4**: 95, 1963.
13. O. Levenspiel, *Chemical Reaction Engineering,* Wiley, New York, 1962.
14. W. F. Hamilton, J. W. Moore, J. M. Kinsman, and R. E. Spurling: Studies on the Circulation, *Am. J. Physiol.* **99**: 534–551, 1932.
15. K. L. Zierler, Circulation Times and the Theory of Indicator-Dilution Methods for Determining Blood Flow and Volume, in *Handbook of Physiology–Circulation,* Washington, D.C., American Physiological Society, 585, 1962, Chapter 18.
16. J. L. Stephenson, Theory of the Measurement of Blood Flow by the Dilution of an Indicator, *Bull. Math. Biophys.* **10**: 117, 1948.
17. D. M. Shames, and P. M. Weber: A General Logical Structure for Quantitative Analysis of Radiocardiographic Data, *Clinical Research* **22**: 210, 1972.
18. Y. Ishii, and W. J. MacIntyre: Measurement of Heart Chamber Values by Analysis of Dilution Curves Simultaneously Recorded by Scintillation Camera, *Circulation* **44**: 37, 1971.
19. E. Rosen, Investigation of Mixing in the Right Atrium, BS Thesis, Columbia University, 1972.
20. D. Parrish, et al.: Analog Computer Analysis of Flow Characteristics and Volume of the Pulmonary Vascular Bed, *Circ. Res.* **7**: 746, 1959.

References

21. P. Wood: Diseases of the Heart and Circulation, Eyre and Spottiswoods, London, 1956.
22. M. L. Marcus, et al.: An Automated Method For the Measurement of Ventricular Volume, *Circulation* **45:** 65–76, 1972.
23. D. Van Dyke, H. O. Anger, R. W. Sullivan, W. R. Vetter, Y. Yano, and H. G. Parker, Cardiac Evaluation from Radioisotope Dynamics, *J. Nucl. Med.,* **13:** 585–592, 1972.
24. W. L. Ashburn, Applications of the Gamma-Ray Scintillation Camera in Cardiovascular Diagnosis, *The Heart Bulletin* **19:** 61–64, 80–81, 1970.
25. L. Rosenthall.: Application of the Gamma-Ray Scintillation Camera to Dynamic Studies in Man, *Radiology* **86:** 634, 1966.

APPROACHES TO MODELING RADIOCARDIOGRAPHIC DATA: COMMENTS ON F. CASTELLANA'S MODELING OF THE CENTRAL CIRCULATION

EDITOR: JAMES B. BASSINGTHWAIGHTE: M.D., PH.D.

Dr. Castellana is tackling a problem that is of fundamental importance to establishing the value of radiocardiography in studies on patients. Many details will have to be worked out before everyone will be satisfied that the technique is exact. Probably it will never be exact, but we will find that imperfect methods are of real diagnostic value. In what follows, I will attempt to suggest general approaches that will need to be simplified or at least rephrased before being totally practical, but which will provoke consideration of some problem areas in the modeling and analysis of radiocardiographic data.

The complex signal that one obtains by external detection over any specified area is a combination of signals from different chambers with differing efficiencies and with time-varying volumes and concentrations. The general equation for a time-varying signal, $S(t)$, obtained via a detector with axial symmetry of its isodose response curves could be written as the integral over the field

$$S(t) = \int_0^x \int_0^y \int_0^z [\alpha(\sqrt{x^2 + y^2}, z)] [C(x,y,z,t)] \, dxdydz$$

where $C(x,y,z,t)$ is the concentration of tracer at time, t, in 3-space; x and y are the horizontal and vertical distances from the axis of symmetry of the collimated detector; z is the distance from the detector (with R and Z large enough to include all regions containing tracer whose emissions might strike the detector); and α is the efficiency of detection of emissions originating at any point in the field of detection. The point to be emphasized is that the signal is an integral in 3-space giving a measure of the total mass of tracer in a rather spread-out volume, and is not at all the same as the concentration at a point in the bloodstream.

An example which is a gross oversimplification may help to make the point. Consider a system consisting of *two linear* components in series, where the detector obtains a signal $S(t)$ proportional to the contents of the second component with equal efficiency from all locations within it (Fig. 1). If an impulse input, an ideal slug injection, is introduced into the entrance, so that $C_{in}(t) = \delta(t)$, then the concentration-time $C_1(t)$, at the entrance to the second component is $h_1(t)$ and at its exit C_{out} is $h_1(t)*h_2(t)$, the convolution of the input $C_1(t)$ with the transport function, $h_2(t)$. Dr. Castellana has considered uniform mixing chambers so that $C_{out}(t)$ is exactly proportional to the mass of tracer in V_2, which is $V_2 C_{out}(t)$, where V_2 is the component volume; with the detector facing exactly the whole of V_2, then $S(t) = \alpha V_2 C_{out}(t)$. The major problem to be worked out in the future

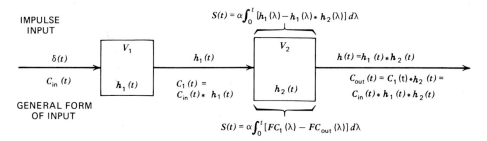

Figure 1. Diagram of 2 component linear system. V_i and $h_i(t)$ are the volume and impulse response of component i. $S(t)$ is the signal proportional to the tracer content of V_2. The equations above the line diagram are associated with an impulse input, those below the diagram with a general input function, $C_{in}(t)$.

is that the $h(t)$'s are not monoexponential, but that each component of the system is distributed. This means that there are finite delays between entrance and exit and that the concentration is not uniform within the volume V_2. The general concept for a steady flow system is given by the equations in the diagram, those above the boxes applying to an impulse input and those below to a general input form. F is flow.

The behavior of the system with the simplest input, a slug injection, is shown in Figure 2, where the impulse responses $h_1(t)$ and $h_2(t)$ characterize the system (1). The total system response through both components is given by $h = h_1 * h_2$ which would give the shape of an output concentration–time curve in response to a slug injection into the input. But the detected signal $S(t)$ has a different shape, being the difference between the integral H_1 and H (defined on the drawing); $S(t)$ has

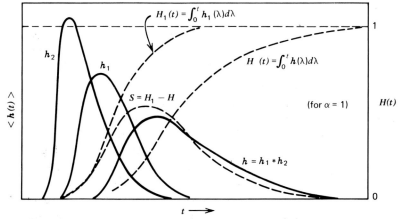

Figure 2. Response of 2-component system to an impulse input $\delta(t)$. (The curves are hand drawn, and not exact.) S or $S(t)$, the thick broken line, represents the tracer content of V_2. This conceptual approach applies whether the V_2 component is lumped or distributed, so long as the detector is sensitive to the whole of V_2.

initially the shape of $H_1(t)$ as tracer enters, but differs as soon as tracer begins to leave the field of the detector (when $h_1 * h_2$ and H begin). Briefly, S lies between H_1 and H and begins to fall when the rate of tracer egress exceeds the rate of tracer entry, and finally takes on the form of $1-H(t)$, which is ordinarily fairly similar to the shape of the tail of the outflow dilution curve $h(t)$. Thus $S(t)$ rises before the outflow curve begins and falls with it.

The implications are several, but the important ones are that the area of $S(t)$ is greater and the mean time shorter than for $h(t)$. If the standard Stewart–Hamilton formulae for flow, F, and mean transit time, $\bar{t}$, were applied using $S(t)$ instead of $h(t)$ or $C_{\text{out}}(t)$, then both F and $\bar{t}$ would be underestimated and the apparent volume, $F\bar{t}$, would be much less than V_2. Since dye-dilution curves are *never* observed to be monoexponential, although parts of their tails may be close, the next stage of development is to consider this more cumbersome but more general approach and to accomplish the difficult task of applying it in the still more complex situations where flows and volumes are inconstant.

In the section Single- and Multiple-Component Models of the Central Circulation, Dr. Castellana quotes Harris' and Newman's categorization of models for components of the circulation as either compartmental or dispersion models. I do not view these as different categories and feel that it may be somewhat misleading to use these words as a basis for classification. In general, the transport functions for compartmental systems are more dispersive than those classified as dispersion models; a first-order mixing chamber produces more dispersion both in space and in the concentration-time curve at the outflow then is present in the tube with fully developed turbulence and high rates of apparent diffusion or dispersion. Furthermore, the two ideas are not separable; for example, consider N first-order mixing chambers in series, which can be called a Poisson model, and note that when N is 30 or greater the model is virtually indistinguishable from a purely Gaussian "dispersion model." I would prefer to stick with the original classification as either descriptive or deterministic. The empirical model that Warner and I chose for providing descriptions of circulatory mass transport, the lagged normal density curve, cannot be classified as either a dispersion or compartmental model since it contains both. It provides an empirical four-parameter fitting function (4) and is useful because it has a wide variety of shapes. A good many other unimodal density functions have similar virtues and will do as well, for example, a random walk with a first traversal (3), log-normal distribution (2). (Figure 3).

My particular objection to first-order compartmental models is that they provide bad descriptions. When coupled with a delay line, which is a model for plug or piston flow, the combination of mixing chamber and delay can give the correct mean transit time for a system and can describe any *one* other higher moment correctly. Usually one would choose to give correct values for first and second moments, the mean transit time, and the variance; then the delayed mixing chamber model *cannot* give correct values for skewness, kurtosis, and so on. Even a single cardiac chamber is not well described by a single uniform mixing chamber; for example, it is clear that the left ventricle does not mix uniformly (9), the last blood entering being among the first to leave, and the blood in the apex is only partly mixed with that in the body of the ventricle or in the outflow tract (7).

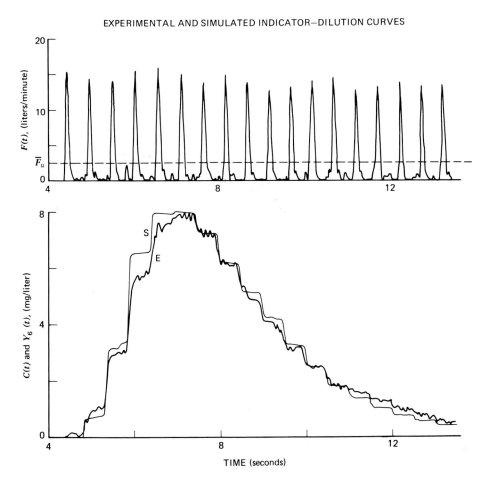

Figure 3. Comparison of model output, $Y_6(t)$, labeled S, with an experimental dye curve from the ascending aorta, $C(t)$, labeled E, when the model was driven by the recorded flowmeter signal $F(t)$. The input pulse to the model was 1 second in duration and was delayed 1.6 seconds compared to the actual injection. The model is an overdamped 6th order operator (defined by 4 parameters, one being proportional to $F(t)$) having a unimodal impulse response. (Taken from (8) with the permission of *Circulation Research*.)

Delay lines are also unrealistic and do not represent intravascular transport even in a cylindrical vessel. Both in the femoral arterial system (4) and the aorta (5), the dye-dilution techniques give an estimate of the average maximum velocity as being 1.65 times the mean velocity in the vessel. This implies a convex velocity profile with a maximum at the axis of the cylinder and the slowest at the wall, the profile being blunter than a parabola. Of course, this average profile is a result also of pulsatile flow and dispersive processes that scatter particles of indicator in both axial and radial directions.

In multipath organs such as the coronary or pulmonary vascular beds, the impulse responses, the probability density function of transit times from inflow to outflow, can also be well described with simple four-parameter descriptive models (6). In a first-order mixing chamber the relative dispersion, the standard deviation divided by the mean transit time, is 1.0. In reality the relative dispersions are much smaller, being 0.2 in arteries of the leg, 0.28 in the aorta, 0.45 across the coronary bed, and about 0.5 through the lung (6).

Dr. Castellana's argument that one should develop a model which realistically takes into account the systolic and diastolic flows through the cardiac chambers makes very good sense. This is an important step. He has shown the advantages of modeling a discontinuous flow, varying-volume situation; to emphasize this he contrasts the solutions with those for a specific constant-flow, constant-volume approximation in his Figure 147. I would argue that there is an intermediate stage which may be useful to consider: the constant-volume, variable-flow analogy, which should be quite applicable in tubular vessels and through the capillary beds of organs. An example of how well this can work for a dilution curve recorded at the root of the aorta can be seen in my Figure 2 taken from reference 8 which shows a reasonable match in the model solution to the step changes in concentration actually recorded there. Since such a model is rather easier to use in certain circumstances than a full-fledged varying-volume, varying-flow model, it may have value for describing these kinds of systems, especially when perfection is not required. Hopefully, as computers get faster, and as more and more information becomes available to permit greater precision in the modeling, fewer simplifications will be necessary. Even while keeping the modeling simple enough that it is manageable, we must continue to strive for conceptual adequacy and to increase realism. There is much to do.

REFERENCES

1. J. B. Bassingthwaighte, H. R. Warner, and E. H. Wood: A Mathematical Description of the Dispersion of Indicator in Blood Traversing an Artery, *Physiologist* **4**(3): 8, 1961.
2. R. W. Stow, and P. S. Hetzel: An Empirical Formula for Indicator-Dilution Curves as Obtained in Human Beings, *J. Appl. Physiol.* **7**(2):161, 1954.
3. C. W. Sheppard: Mathematical Considerations of Indicator-Dilution Techniques, *Minnesota Med.* **37**: 93, 1954.
4. J. B. Bassingthwaighte: Plasma Indicator Dispersion in Arteries of the Human Leg, *Circ. Res.* **19**: 332, 1966.
5. J. B. Bassingthwaighte, and F. H. Ackerman: Mathematical Linearity of Circulatory Transport, *J. Appl. Physiol.* **22**: 879–888, 1967.
6. T. J. Knopp, and J. B. Bassingthwaighte: Effect of Flow on Transpulmonary Circulatory Transport Functions, *J. Appl. Physiol.* **27**:(1): 36, 1969.
7. L. D. Homer, and H. P. Krayenbuehl: A Mathematical Model for the Estimation of Heart Volumes from Indicator-Dilution Curves, *Circulation Res.* **20**: 299–305, 1967.
8. J. B. Bassingthwaighte, T. J. Knopp, and D. U. Anderson: Flow Estimation by Indicator Dilution (Bolus Injection): Reduction of Errors due to Time-Averaged Sampling During Unsteady Flow, *Circulation Res.* **27**: 277, 1970.
9. H. Irisawa, M. F. Wilson, and R. F. Rushmer: Left Ventricle as a Mixing Chamber, *Circulation Res.* **8**: 183–187, 1960.

PRINCIPAL EDITOR: R. H. JONES

CONTRIBUTING EDITORS: LEON LIDOFSKY, THOMAS BUDINGER, AND PAUL WEBER

8. Instrumentation

The clinical accuracy of radionuclide cardiac studies depends upon the initial quality of the observed and recorded data, which is determined in large part by the performance of instruments used. Whereas details of instrumentation need concern only the physician responsible for data acquisition, an understanding of the basic approaches to radionuclide detection is essential to any clinician who wishes to evaluate the accuracy and reliability of cardiac studies. This chapter surveys the major considerations of instrument performance for radionuclide cardiac studies and supplies sufficient details of equipment used in studies reported in this book to repeat the techniques described. General considerations of radionuclide detection instrumentation have been thoroughly reviewed in a number of textbooks and articles and reiteration of these principles is beyond the scope of this contribution. The aspects of instrument performance pertinent to cardiac studies are emphasized with no intention to detract from the importance of the basic principles of radionuclide detection. The major unique feature of cardiac studies results from the rapidity of blood flow through the central circulation which limits the total counting time available to less than 20 seconds. If count changes with each cardiac contraction are to be indexed, individual counting intervals must be no greater than 0.1 second. The fine spatial and temporal resolution necessary for optimal cardiac studies stresses the capabilities of all currently available instrumentation. Furthermore, the resulting mass of data presents a formidable challenge to accurately store, recall, and analyze.

The first radionuclide cardiac transit studies were performed using single Geiger-Muller tubes or scintillation probe systems interfaced to a strip-chart recorder. These relatively inexpensive and simple devices can be easily taken to the patient bedside. Time–activity curves representing all counts within the collimator field result from single-probe studies. The utility of the resulting data depends largely upon the accuracy of anatomic positioning of the probe. Careful positioning of the probe over rather large anatomic areas such as the lung or left ventricle provides reproducible data. However, an abnormal cardiac configuration may result in data which are difficult to assess. Also, anatomic regions smaller in size than the field of view of the probe cannot be accurately separated from adjacent structures.

The gamma-camera holds one major advantage over a single probe for cardiac studies in that spatial orientation of counts is retained over a large area. Therefore, the resulting data can be presented in pictorial form which permits appreciation of the sequential passage of a radionuclide bolus through the central circulation. Alternately, quantitative data can be obtained from any number of subregions of the detector field selected to correspond to specific cardiac chambers. The large volume of data produced during camera cardiac studies can be practically utilized only with a recording system which permits retrieval and manipulation of data after completion of the study. The simplest data retrieval equipment uses video instrumentation for storing and processing images. A computer interfaced to the gamma-camera has proved a useful method for retrieving quantitative data.

Two different types of gamma-cameras are currently available. A multicrystal camera such as the Baird-Atomic System 70 utilizes a matrix of individual scintillation crystals with digital recording of the output of each. This instrument has the potential advantage of a higher counting rate than a single-crystal device, but for dynamic cardiac studies has intrinsic resolution limited by the size of individual crystals 1cm × 1cm in counting area. The single-crystal cameras based on the principle described by Anger utilize electronic positioning of interactions occurring within a single large scintillation crystal. Single-crystal instruments have become the most widely used gamma-cameras in nuclear medicine and remain the most frequently used instrument for cardiac studies. No single instrument currently available demonstrates performance characteristics ideal for all cardiac studies. However, an understanding of the advantages and limitations of equipment now available aids the choice of instrumention for specific procedures.

VIDEO INSTRUMENTATION FOR THE GAMMA-CAMERA

In 1969 Kriss, Bonner, and Levinthal described a system, the variable time-lapse videoscintiscope (VTV), which adapted the scintillation camera for radionuclide angiocardiography (1). The components of the VTV system include a television camera mounted to view the face of the auxiliary cathode ray tube of the scintillation camera (Pho-Gamma III, Nuclear Chicago, Inc.). The video signals from the camera and an audio signal are recorded on video tape. The audio signal activates a timer which triggers an exposure after a selected interval as short as 1/60 second or as long as 16.7 seconds. Vertical synchronizing signals from a television monitor regulate the interval timer to insure precise matching of the video image with actual elapsed time. When the preset time of the interval timer has elapsed, the exposure timer is activated and simultaneously the television monitor screen is illuminated. The screen remains illuminated only during the interval preselected on the exposure timer (as short as 1/60 second to as long as 16.7 seconds). The image on the screen is photographed by a Polaroid camera. Brightness and contrast controls on the monitor let the viewer choose the optimum optical conditions for obtaining photographic images. The photographs can be trimmed and mounted in a desired sequence for further study and analysis.

Another analog system which permits storage of radionuclide angiogram data is the rotating memory videoscintiscope (RMV) (VAS, Ltd., Sunnyvale, Calif.). The RMV is a compact, versatile, and easily operated device which has been developed for recording and documenting dynamic events sensed by a scintillation camera. The X, Y, and Z outputs of the camera are transformed into standard television format by a scan converter which provides electronically controlled integration of the information before recording takes place. The video images are simultaneously displayed on a television monitor and recorded within a memory which records the signals on a magnetic disc rotating at 3600 rpm. The recorder has a capacity of 570 images. Controls enable adjustment of scintillation intensity, spot size, spot focusing, and recording rate. The operator may vary the time for integration of scintillations from 0.1 to 5 seconds. Regardless of the rate of recording, no information is lost in the recording process.

After a study has been recorded, it may be replayed into a television monitor as a series of rapidly or slowly displayed integrated images depending on the setting of a playback speed control. Depending on the rate of recording, the rate of playback can be varied over a continuous range from "freeze" to 300× real time. Each image is identified with a displayed number, providing ease of starting and stopping any desired sequence. Simple controls enable the operator to advance or retreat one frame at a time. Any single image or sequences of images can be photographed by a Polaroid camera bracketed to the television monitor. Summation controls permit the automatic sequencing of any preselected series of images while the photographic camera is being exposed. This permits an integration of scintillation information over any desired portion of the study.

Components which enable quantitative information to be obtained simultaneously from two selected areas of interest consist of dual video scintillation counters and a dual strip chart recorder. Paired vertical cursors may be positioned on the television screen after completion of a study. Playback of data then generates on the strip-recorder the count rates from the rectangular regions of interest defined between each set of parallel cursors. Controls on the video scintillation counters permit choice of length of each cursor bar, the width between a cursor pair, and their general position on the screen. Analog time–activity curves may be generated from different portions of the heart as the radioactive bolus traverses the cardiac chambers. However, careful calibration of the system is required to attain the necessary linear relationship between count rate and the output of the cathode ray tube if quantitative analysis of indicator-dilution curves is intended. This added quantitative feature, while providing very useful information, has proved somewhat cumbersome in practice due to (a) the restriction to a rectangular format and vertical position, (b) the few areas which can be studied simultaneously, (c) difficulty in standardization, and (d) lack of computational capability. Most workers interested in quantitative analysis of angiographic data prefer use of digital computer systems, which offer computational precision but are considerably more expensive and which, in most instances, are associated with a much poorer quality scintiphotographic image than can be generated with the analog system.

THE SCINTILLATION CAMERA TO COMPUTER INTERFACE

The interfacing of a small digital computer to a single-crystal gamma-camera provides a system which may be used for quantitative radionuclide angiocardiography. These computer systems involve components responsible for acquisition, storage, display, processing, and hard-copy output of data. These components are considered separately in this report, with emphasis on the most stringent functional requirements for each category.

Data Acquisition

Ideal data acquisition provides information with good spatial resolution at high counting rates. The maximum spatial resolution of the scintillation camera using a high-resolution parallel-hole collimator is about 10 mm at the collimator surface, and less at the depth of the heart and great vessels. The spatial resolution present in a digitized image—that is, one in which analog-to-digital converters have placed each scintillation recorded within the crystal into a space matrix within the computer core memory—depends upon the maximal resolution frequency across the crystal surface. Since information theory (2) requires two digital points for every resolution cycle, for maximal resolution of 8 mm

$$\frac{254 \text{ mm crystal diameter}}{8 \text{ mm maximal resolution}} = 31.75 \text{ resolution cycles across the crystal} \times 2 = 63.$$

Thus 63 sample points are required so that a space matrix of 64 × 64 elements is adequate. If resolution within the crystal could be improved to 5 mm, 102 sample points would be required to avoid image deterioration and a matrix of 128 × 128 elements would then be necessary. Resolution at 8 mm in a 64 × 64 matrix can be obtained with computers using either 12-bit or 16-bit words, but 5-mm resolution would require data to be acquired in a list mode format, a list of consecutively occurring x and y addresses with time markers.

In adults, the heart chambers, great vessels, and lungs are sufficiently large and discrete to permit excellent clarity with a space matrix of 4096 elements (64 × 64) and adequate resolution for many studies is obtained with a 32 × 32 matrix. In children, the small size of the heart and great vessels in relation to the crystal detector makes it much more difficult to identify the individual cardiac chambers. Although maximal spatial resolution of the camera remains fixed, digital resolution can be improved toward that limit by use of zooming analog-to-digital converters. These devices alter the gain so that only a small portion of the field of the scintillation crystal fills the entire matrix chosen.

"Temporal resolution" can be defined as use of the time index during which data are acquired in order to identify an anatomic or physiologic entity. A selected precisely timed space matrix is called a "time frame." The specific time frequency response required depends upon the type of dynamic data analysis contemplated. The highest framing rate or greatest frequency response presently anticipated is that required to measure the ejection fraction of the left ventricle, and particularly to obtain an accurate curve of ventricular activity (volume) as it changes with time (dV/dt). Chapter 5 contains explicit definition of this function and its derivation. A minimum of 10 data points per cardiac cycle may be assumed to be necessary to define this curve, so long as it contains no sudden inflections which must be accurately measured. The limits within which this assumption is justified are discussed in Chapter 5 and shown in Figure 94. If this be true, a heart rate of 60 beats/minute requires a minimum framing rate of 10 frames/second, while a heart rate of 180 beats/minute requires 30 frames/second. In the most stringent situation, neonatal angiocardiography, the maximum framing rate required is no more than 50 time frames/second, or 20 msec for each time frame. However, a medium frequency response is adequate for much hemodynamic information. A time frame duration of 0.5 or 1.0 second is suitable for calculating cardiac output (3), evaluating left-to-right shunting from pulmonary washout curves (4), performing transit time analyses (5), deconvolution, and generating coronary washout curves.

In cardiac studies with the requirement for fine spatial and high temporal resolution, data density must be high to permit statistically valid conclusions. When there is a high rate of input into the system, data loss may occur, either in the scintillation camera or in the computer system. Data loss rates are complex functions which are conveniently expressed as the dead-time of the whole system, and may be determined by experimental measurement when a complete system is assembled. The dead-time of the computer system results primarily from the processing time of relatively slow (50–100 MHz) analog-to-digital converters (ADC). The contribution of the computer to the total dead-time can therefore be reduced by using higher frequency analog-to-digital converters (200 MHz) and/or by buffering and derandomizing the 50 to 100 MHz ADCs. Computer systems may be interfaced with a scintillation camera capable of handling more than 40,000 cps without significant loss of computer-processed data, but 5 μsec camera dead-time results in a 10% data loss at this count rate and becomes the limiting factor. The relationship between count rate, camera dead-time, and data loss has been calculated by Budinger (Fig. 150).

Data Storage

Initially, acquired data enter the computer core or an analagous memory core that is inadequate for long-term storage. The usual technique for transferring data to another storage device uses two portions of computer core for data acquisition.

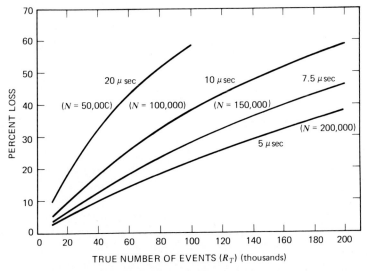

Fig. 150. These data illustrate the percent count loss as a function of the true counting rate for system dead-times of 20, 10, 7.5, and 5 μsec. The theoretic maximum counts per second (N) which can be achieved at each dead-time is also demonstrated.

This permits data in one block of core to be transferred to mass storage, while data are simultaneously being acquired in the other block of core. By such flip-flopping, data may be acquired without loss.

The two types of storage devices in common use are magnetic discs and incremental magnetic tape. Discs are more expensive and have smaller capacity than tape, but data on discs are rapidly accessible for retrieval and processing. They can be transferred in blocks from computer core to storage at rates as high as six to twenty 4000 word frames of data per second. Magnetic incremental tape recorders provide much less expensive data storage, but their transfer rate from computer core is usually no faster than 3 frames/second. The accessibility of tape-stored data is poor because the access is sequential rather than random. In order to find the proper record or sequence of records, the tape recorder must laboriously search backward and forward through the length of the tape on which the data are stored. New phase mode or double-density tape recorders, though not generally available commercially at this time, can transfer from core up to eight 4000 word blocks per second.

Data can be stored either formatted as time frames (histogram mode) or as individual events (list mode). When formatted as time frames, the matrix parameters are preprogrammed before data acquisition. Acquired data are then transferred to the storage device in the form of matrix time frames (Fig. 151). When formatted in list mode, sequential locations of computer core contain x,y position coordinate information as two 6-bit, two 7-bit, or two 8-bit words (Fig. 152). These coordinates separated by timing pulses at regular intervals are then transferred to storage for off-line playback and reformatting as time frames. Six-bit ADCs permit 64×64

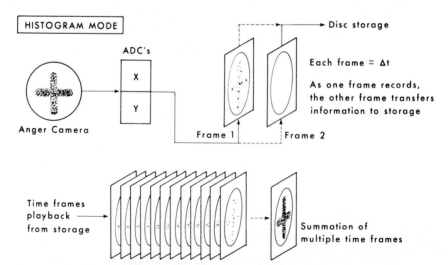

Fig. 151. Schematic representation of histogram mode acquisition format showing several successive time frames. Data are acquired for a specific time period for each frame with the spatial distribution of the activity within the x, y space matrix corresponding to the location of each scintillation crystal. If more than one event occurs within a given matrix location during a particular time period it is visualized on the display of the matrix as an increase in intensity of the dot. Each matrix location has a maximum capacity for events dependent upon the bit size of the computer. A 12-bit computer could store 4096 counts in each matrix location. The last time frame represents the summed integration of the earlier individual time frames.

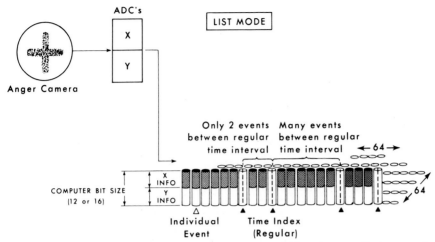

Fig. 152. Schematic representation of list mode acquisition format. Each computer core location sequentially contains x, y position coordinate information as two individual words depending upon the bit size of the computer core. Timing pulses at regular intervals separate this coordinate information. No image is visible directly from the list mode format without reformatting as histogram time frames.

data point spatial resolution; with 8-bits, resolution can be increased to 256×256. With a 16-bit computer a single word can be used for an 8-bit x coordinate and an 8-bit y coordinate, while with a 12-bit computer double precision is required with alternate words used for the x and y coordinates. Use of 8-bit format is much less conservative of core space and therefore requires a faster transfer rate to storage for the same data input rate than with a 16-bit machine.

The list mode of data acquisition also permits framing rates to be determined after the study has been completed and can produce framing rates of a much higher frequency that the histogram mode. Since in list mode final formatting is done after completion of the study, only the frequency of the timing signal limits the framing rate that can be selected. A disadvantage of list mode is that the acquisition data rate is limited to that which can be transferred in 4000-word blocks to the storage device. Thus, in many small computers a maximum transfer rate of 6 frames/second would limit the input count rate to 24,000 cps. The more rapid rate of data transfer from computer core to disc is in part offset by limited disc capacity. This capacity may be exceeded during a high-frequency study performed in either histogram or list mode at high framing rates for longer than 30 seconds. After the injection of a radioactive bolus the first pass through the central circulation is usually completed within 30 seconds. However, the capacity for continuous acquisition for 1 minute, which is available in tape, may be advantageous. If possible, it is advisable to have more than one mass data storage device. One or more discs for rapid data acquisition and processing, coupled with incremental magnetic tape for long-term permanent storage, would be the most economical option.

Data Display

On-line digital data display is usually not necessary. Two aspects of off-line data display must be considered. First, the data must be retrievable for display on an oscilloscope so that the study can be reviewed sequentially in part or in whole. This permits processing the images for anatomic detail so that appropriate areas of interest may be flagged for histogram generation. Interaction with the actual data usually consists either in flagging areas of interest with a light pen, or defining areas of interest by means of key controls that move lines, windows, or points over the surface of the oscilloscope. In addition, there must be some form of display in association with the control terminal, either teletype paper or, preferably, a cathode ray tube, so that particular interventions can be recorded and the output of data from such interaction may be made visible.

Specific Components for Assembling a Computer System

Although the nature of the cardiac study planned defines the characteristics required of the instrumentation used, an idealized system may be described which would prove suitable for most cardiac studies (Fig. 153). Two analog-to-digital converters digitize the x and y signals from the camera. The frequency response of devices commercially available varies from 50 to 200 MHz. Two-hundred-megahertz ADCs have a pulse pair resolution of 3 μsec and add no significant time loss to the 5-μsec camera dead-time. Computer dead-time with 50-MHz ADCs can be diminished by analog buffering and derandomizing the input to the ADCs, and by digital buffering of their output. Nevertheless, most computer systems marketed for high-frequency studies today do add significant dead-time to that of presently available cameras. When dead-time of the computer system is a problem, it can be significantly decreased if data acquisition is buffered by recording the output of the scintillation camera directly on magnetic tape with an analog tape recorder. Four-channel frequency-modulated tape recorders are available which can be interfaced

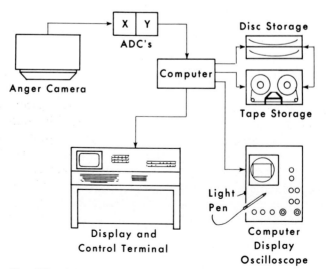

Fig. 153. Schematic representation of individual components of a computer system interfaced to a scintillation camera showing their functional relationships.

directly to the scintillation camera so that the outputs of the x and y signals are each recorded on their respective channels with a data loss of less than 1% at count rates up to 28,000 cps and with spatial resolution also maintained within 1%. The other two channels may be used for voice and an analog (such as EKG or pulse) waveform. During acquisition, the tape speed for recording can be up to 60 in./second for maximal data acquisiton rates. The tape can then be slowed to $3\frac{3}{4}$ in./ second for off-line playback into the computer system for digital processing. The 16-fold reduction in speed acts as an excellent buffer, permitting a 16-fold increase in framing rate without data loss even with unbuffered 50-MHz ADCs. Such an analog buffer tape thus becomes analogous to list mode storage in computer core memory in that the data are not time-frame formatted until after acquisition. In addition, FM tape has the virtues of large-capacity and low-cost permanent storage.

Some minicomputers available today have word lengths of 12-bits and some of 16-bits. The latter are newer, and data acquired on these 16-bit computers are more directly compatible for inputting into large computers. One 16-bit computer now available is byte oriented. This means that the 16-bit words can be treated as two 8-bit words, thus doubling the number of time-frames of a given matrix size. This in turn doubles the maximum framing rate though with a data density limit of 8-bits or 256 counts per address. Cycle times are short, ranging from less than 2 μsec, differences which are not of significance in physiological studies. Only moving-head discs are used in these systems. Double-density discs, which double the disc capacity, have recently become available. Fixed-head discs can have a faster transfer time, but they are much more expensive and have not been incorporated into present systems. Only one manufacturer offers a disc system capable of transferring 20 4K blocks of core data per second. All others are limited to 6 or 7 such transfers per second.

Both refresh and storage display oscilloscopes are available. Refresh-type tubes must have the data generated from memory core at a frequency rate at least 25 frames/second in order to maintain a flicker-free image on the scope. Intensity modulation achieves gray scale effects. This refresh requirement makes difficult the display of alpha-numeric data with the image. Storage oscilloscopes do not require refreshing and can therefore be used to display alpha-numeric data with the image, with less processer time and without flicker. Display and decay times are much longer with storage oscilloscopes, and rapidly displayed sequential images giving a motion-picturelike effect are not possible with a storage scope. An electric control terminal with oscilloscope display provides easier input and output than the conventional teletype terminal. The oscilloscope display of alpha-numerics becomes important when long digital readouts are required because of the slow output of the conventional teletype.

Two reasonable approaches are available for those interested in assembling a system today. One may purchase a predesigned system with suitable specifications from a given manufacturer. This approach has the virtue of defining the number of options available, committing system maintenance responsibility to a single supplier, acquiring a system that should function well when installed with no additional interfacing costs, and perhaps most important of all, acquiring the relatively complete software package that will come with the prepackaged hardware. The software should provide all necessary routine functions including the ability to acquire, recall, format, and transfer data; perform simple mathematical functions upon data; generate histograms of flagged areas of interest; smooth data; and correct for data rate nonlinearity, crystal nonuniformity, and data loss. The dollar value of this software package is difficult to estimate, but it would certainly be optimistic to suggest that it could be developed in the laboratory within 6 to 12 months after arrival of the hardware. The second alternative is to select individual components, interface them to each other and to the scintillation camera in the laboratory, then write all software. If this approach is taken, one must have electronics engineering personnel to assume responsibility for maintenance and interfacing, and one must have extensive capacity for software development.

To purchase a predesigned system thus has obvious advantages over the second alternative. Nevertheless, most manufacturer-designed packages do not arrive in the laboratory totally trouble-free. A shake-down period is usually required to work out problems. Some software must be developed for special operations. For any system, performance parameters including dead-time, frequency response, and maximal spatial resolution should be measured before the system is used clinically.

DATA CORRECTION FOR INSTRUMENTAL ABERRATIONS

Although the counting rate loss is Poisson distributed, neither the instantaneous rates nor the exact effects of dead-time are known *a priori*. However, several schemes are satisfactory for dead-time-correction. First, a fixed long-lived source may be attached to the gamma-camera at a position outside the area of interest. The count from that source should be constant. Each frame counted for a fixed time can be normalized to the counts in that area. Since counting rates change slowly relative to the frame time, good counting statistics may be developed by averaging that count over many frames for use as reference. Alternatively, pulses from a pulser of known, fixed rate may be injected into the data scheme if, at their time of generation, the system is not "busy." Finally, if the incident energy spectrum is time independent and the losses low (10–15%), the following procedure would suffice: use the system to measure the rate from a decaying source of the isotope of interest over several half-lives and derive a correction factor versus observed count rate. Then calculate the observed rate for each frame of interest and apply the correction. The problem of spatial distortion is much more subtle and dependent on the individual instrument and its state of adjustment.

The correction procedure for the mosaic camera could be accomplished by recording a "flood" exposure and storing it. Then, the data for each element are normalized by the count in the corresponding element under flood exposure. The procedure for the Anger camera is much more subtle in that the correction depends on the absolute position of the crystal and varies in an irregular manner. However, for most applications the requirement is *not* to obtain an undistorted picture. Rather, it is to attempt to establish a one-to-one correspondence between a position on the crystal and a position within a compartment of the cardiovascular system. From this point of view *no* correction for nonlinearity is necessary. Sharpening of images for either camera can sometimes be attained by appropriate transformation of the data using Hadamard or Fourier transforms. Under any circumstances the procedure will serve to smooth the data from statistical fluctuations.

Definition of Anatomic Regions

Traditionally, an "on-line" physician has served to designate regions on the image plane which are to be considered to relate to a specific chamber within the heart–lung system. It is possible to ease this interaction by providing hardware devices which will allow him to work directly from the graphic image without the necessity of typing, punching cards, or engaging in any other similar digitization. It would relieve considerable load on the physician if the regions could be determined with minimal or even no direct attention on his part. Some approaches can be described as follows.

On-Line Human Control

The time-sequenced outputs of histograms formed from camera data may be displayed on a display terminal. If the physician is to define regions, it is possible that he will wish histogram time frames somewhat longer than those to be used for ultimate reduction. Thus, some integration is desirable. A storage oscilloscope or variable-persistence oscilloscope may be used, although somewhat more flexibility is possible if the data is integrated by the computer and the integrated results presented. When a frame with recognizable outlines has been generated, the physician will wish to designate regions of interest. The appropriate procedure is strongly influenced by the particular display device he is using.

Storage or Variable Persistence Oscilloscope

Some oscilloscopes (TEKTRONIX 611 for instance) allow points to be displayed in both storage and nonstorage mode. The operator may then use a device which will cause its position as moved by his hand (tablet, sonic pen, joystick, mouse) to generate a nonstored image superposed on the stored image of the oscilloscope. Whenever he has determined a point on an outline which he deems satisfactory, he may operate a switch to store it permanently. He repeats this procedure until he has defined a boundary for the region he wishes to select. The selection may be point-by-point, or line segment by line segment, again depending on the particular hardware available.

If the graphic device is a conventional X–Y oscilloscope with image refreshed from a list of points stored in the computer, the devices just suggested may be used. However, an additional device, a light pen, offers some advantage. The histogram display is an array of points each corresponding to a number stored in memory. The user may designate any point on the display by pointing the light pen (a photosensitive device) at that point. When it is displayed, the light pen detects the light flash, and emits a signal which allows the computer to label the storage location corresponding to the point selected as interesting. Usually the program will alter the display to allow the operator to recognize those points he has selected (by change of intensity). When he has selected all points corresponding to his selected region, he operates a switch to signal the computer to stop the procedure.

Either of these methods is satisfactory if one accepts the premise that the operator is available and is correct in his designations. Some methods have been used and others may be suggested which may serve to supplement his decisions by using them as an initial approach to a less subjective procedure or to replace him entirely. It has not yet been determined if the second level is possible.

Automatic Definition of Regions of Interest (ROI)

The chambers of the heart–lung system are considered to be well mixed at all times. Thus the time behavior of isotope concentration at any point within a chamber is the same as that for any other point within the same chamber. This oversimplified concept is the basis for automatic ROI definition. The oversimplification comes because it is mass of isotope (concentration x chamber depth) which is observed and because it is difficult to avoid overlap of chambers. The first of these difficulties can be minimized by timing with respect to a heart-based physiological clock (EKG), while the second can be minimized by using a preliminary selection by the physician.

Jones et al have defined correlations in terms of selected attributes of the entire distribution associated with each counting cell, in particular, time of maximum count and semilogarithmic slope of curve downsweep. They have chosen to smooth data by Fourier analysis. However, their reports have indicated that the magnitude of computation required (at that time—1972) the use of a large offline computer system. For this reason other attributes, such as time-moments of the distribution, should be considered as not requiring smoothing and, hence, being less demanding of computer facilities.

The possibility of rather more complete analysis than is now carried out exists because counts at a point come from a fixed (and small) number of chambers. Thus a procedure should be considered in which the most clearly distinguished regions are designated and analyzed. When regions corresponding to a single chamber have been defined and the parameters used for selection are also defined, an iterative procedure in which known regions are "peeled off" to uncover regions with two-chamber overlap may be pursued to ultimately recover the behavior of the particular chambers of greatest interest.

Procedures described above have been used in other fields for deconvolving images and for separating single components from sums of components. It is probable, at least for the present, that the magnitude of data (up to 128×128 channels and five or six components) may require the use of large computers for their solution.

PROPERTIES OF MINICOMPUTERS

The definition of a "minicomputer system" is largely subjective. A pragmatic definition is: "A short word-length computer system whose cost is sufficiently low that its exclusive use for a special purpose can be justified." While exceptions can be found, this emphasizes a primary factor—cost. The revolution in increasing power, speed, and size coupled with decreasing price shows no sign of abating. Thus definitions based on absolute size or cost are rather temporal.

Those minimum properties necessary for applicability in radiocardiography include:

a. Instruction processing time $\lesssim$ several μsec.
b. Memory speed $\lesssim$ 1 μsec.
c. At least two data channels.
d. Word length at least 16-bits.
e. At least one bulk storage device capable of effective transfer rates for large blocks of data $\gtrsim$ 60 kbyte/second.
f. Bulk storage device with capacity $\gtrsim$ 10^6 word.
g. Industry-compatible magnetic tape (if not already present as f).

Large versus Small Computers

In the past several years the distinctions between large and small computer systems have changed. Rather than cast the choice in such general terms, it is preferable to examine the requirements placed by the measurement and analysis procedures on the instruments needed to carry them out.

During data acquisition, highest priority must be placed on transferral of individual descriptors to buffer storage. When a buffer is full, a new buffer must be assigned and the previous data stored or added to a histogram to be stored at later time. Such a requirement for priority is automatically satisfied in a small, single-user system. It is less likely to be possible in a large, multiuser environment.

Data reduction and final analysis place considerably less emphasis on priority, but do stress the need for: (*1*) adequate core storage to process a meaningful quantity of data and (*2*) sufficient hardware power to carry out the necessary calculations in an acceptably short time. Very small systems (<8K core) will almost certainly be too small for data reduction and will certainly be too small for any but the simplest data analysis. However, "small" systems are now economically feasible (>16K core, disc operating system) which approach the capabilities of some large computers of less than a decade ago. They are limited primarily only in the solution of very large, curve-fitting analysis programs in that their speed is one twentieth to one thousandth that of some modern computers for real arithmetic procedures and their core limits them to small arrays (<4000 numbers). Their hardware power is useful only if an adequate operating system is available. This system must allow the use of disc random-access files in a compiler language environment. Typically FORTRAN IV is the compiler language of choice, primarily because it is so widely used. Many standard programs are available in FORTRAN IV and may be adapted with minimum effort. Languages such as BASIC are also useful, particularly in developing data reduction programs. The lack of standardization and the smaller population of users tends to limit program exchangeability.

If convenient access to a large computer system is available, it may be preferable to restrict the dedicated device to the smallest system which will take data at the required rate and with sufficient bulk storage. In this case, communication between that device and the large computer is essential, with industry-standard tape the medium of choice. It is worth noting that the fractional difference in price between a system which will take data at high rates and with large storage capacity and one which will also carry out considerable analysis is not very large.

CONCLUSION

Today it is both economically and technically feasible to interface a small computer to an Anger-type scintillation camera. The addition of a computer greatly expands the scope and utility of the scintillation camera and makes it possible to perform many studies involving the heart and central circulation that could not otherwise be performed. The future will bring computers with faster cycle times, less expensive core memories, more sophisticated peripheral devices, and less expensive disc and tape recorder packages, all resulting in making these devices available to more institutions.

HIGH COUNTING RATE PERFORMANCE OF THE ANGER SCINTILLATION CAMERA

The dynamics of the cardiovascular system reflected by radionuclide flow through the heart can be described by sequential images obtained as rapidly as 50/second. Quantitative information regarding shunts, left ventricular performance, and anatomical abnormalities can be extracted if the statistics and spatial resolution are adequate. This report examines the potentials and limitations of the scintillation camera for rapid cardiac studies and the influence of count rate on spatial resolution. The object is to determine the count rate above which pulse pileup in the scintillation camera electronics limits resolution and contrast.

Methods

Resolution versus counting-rate studies were made using bar patterns with bars $\frac{3}{16}$, $\frac{1}{4}$, $\frac{1}{2}$, and $\frac{9}{16}$ -in. (4.8, 6.4, 12.7, and 14.3 mm) apart (Fig. 154). In the counting rate versus source-strength studies the bar pattern was placed over the uncollimated crystal, and calibrated point sources between 0.05 and 6.6 mCi were used at a distance of 48 ± 2 in. from the crystal. The sources consisted of ^{99m}Tc calibrated at the time of measurement. The observed counting rate was determined by measuring the time required to gather 999,000 accepted events. The source strength was related to true counting rate by calculation of the expected true counting rate using

$$R_t = \frac{R_0}{1 - R_0 t}$$

once t was determined from

$$t = \frac{R_{12} - R_1 R_2 (R_{12} - R_1)(R_{12} - R_2)}{R_1 R_2 R_{12}}$$

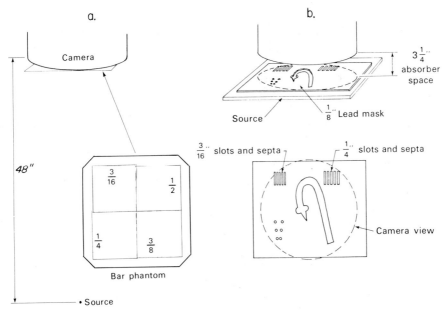

Fig. 154. This experimental setup was used to assess the spatial resolution of the gamma camera as a function of count rate.

where R_1 and R_2 are two different counting rates associated with two sources, and R_{12} is the observed rate from the two sources counted simultaneously. The alternate method of establishing the expected counting rate consists of establishing the slope of a straight line which passes through the low counting rate points. The two methods give the same intercept.

The effect of absorber on resolution and count rate was assessed using a $\frac{1}{8}$-in. mask inserted over a distributed ^{99m}Tc source in a Lucite container. This transmission phantom was positioned at various distances from the collimated crystal in air and masonite absorber to simulate isotope distributions similar to that found in a cardiac flow study.

Results

The counting-rate capabilities for the Nuclear Chicago Pho Gamma III, Nuclear Chicago High Performance, and Picker 2C cameras are shown in Fig. 155. These data were obtained on two or more cameras of the same type and the results did not vary significantly. The Nuclear Chicago Pho Gamma III saturates at 80,000 to 100,000 cps, and the Nuclear Chicago High Performance camera saturates at 50,000 to 60,000 cps. These counting rates are achieved at rates of 150,000 to 200,000 cps within the photopeak and at a gross event rate on the crystal of 400,000 cps. The Picker cameras have a low counting rate performance compared to the Nuclear Chicago cameras, with a maximum counting rate near 24,000 cps. However, the Picker 2C camera maintains high resolution and acquires little spatial distortion at and beyond the point of maximum observed counting rate.

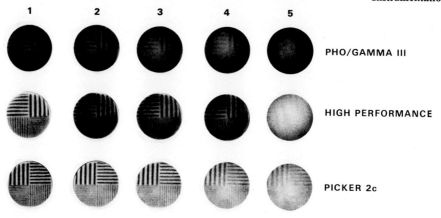

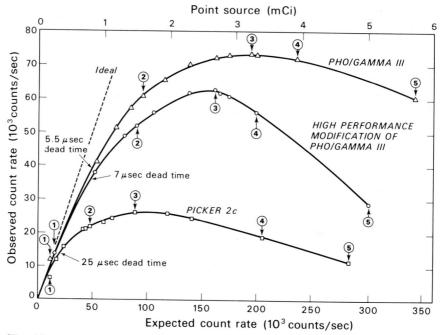

Fig. 155. Data demonstrating the count rates observed with different instruments with increasing radioactivity. Images obtained at different counting rates depicted on the curve relate the imaging capabilities of these systems at increasing count rates.

A comparison of resolution and spatial distortion for the three classes of cameras (Fig. 155) reveals the Picker camera to give superior results at expected event rates of 100,000 to 200,000 cps. This superiority is at the expense of relatively high data loss.

The effect of window-width change is shown in Fig. 156, and the comparison between the Picker 2 and Picker 2C cameras in Fig. 157. The difference between the two Picker cameras is thought to be due to a difference in window setting between the two experiments. The resolution of the Picker 2C (Fig. 155) was superior to the Picker 2.

Performance with Tissue Absorbers

There is little loss in resolution as counting rate is increased to the maximum performance rate of the Picker 2 or 2C cameras and the Nuclear Chicago High Performance cameras. However, tissue absorption and scattering, resulting in more Compton-degraded photons falling outside the energy window, might alter resolution differently with increasing count rates. Using the phantom of Fig. 154, the effects of counting rate and tissue absorption on resolution were measured.

With the transmission phantom at the collimator there was little significant deterioration in resolution as the counting rate was increased to 50,000 cps. Results at 50,000 cps (accepted events) were similar to those at 5,000 and 35,000 cps. shown in Fig. 158. Holes or bars $\frac{3}{16}$-in. wide separated by $\frac{3}{16}$ in. can be visualized. The high-resolution collimator and Nuclear Chicago High Performance camera were used.

With $1\frac{1}{2}$ in. of masonite absorber (sp gr 1.1) interspersed between the source and the collimator, the $\frac{3}{16}$-in. resolution is lost, but the $\frac{1}{4}$-in. resolution remains up to and beyond 45,000 cps.

With 3 in. of absorber interspersed between the phantom and the camera, the $\frac{1}{4}$-in. bars are barely visible. The resolution has deteriorated to between $\frac{1}{4}$ and $\frac{7}{16}$ in. (separation of middle set of point sources). This serious deterioration of resolution is based more on the 3-in. distance than the presence of absorber, since resolution deterioration with absorber was indistinguishable from that found in air.

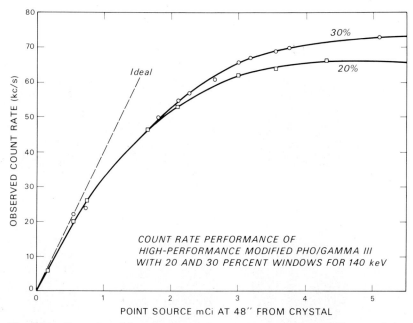

Fig. 156. Data describing the observed count rates with two different window settings illustrate that a higher level of camera saturation is achieved with the broader window.

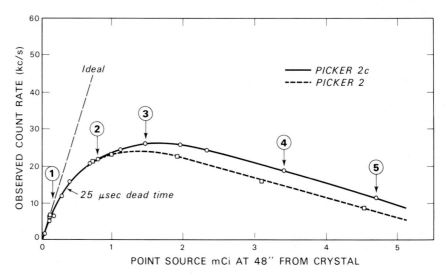

Fig. 157. Data comparing the count rate performance of the Picker 2 and Picker 2C cameras.

Conclusions

Radionuclide cardiac studies involving intravenous or central venous injection of 10–15 mCi ^{99m}Tc will give observed counting rates of 30,000–40,000 cps during a few seconds of the flow study with no loss in spatial resolution if the camera has appropriate high-speed electronics. The basic limitation in attaining sufficient statistics for the determination and quantitation of cardiac flow patterns is the speed

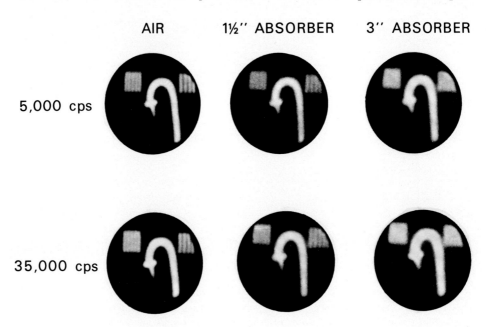

Fig. 158. These images demonstrate the similarity of images obtained at two different counting rates with increased scatter medium.

of the electronic components. The resolution stability with increasing counting rates depends on the light-gathering geometry, photomultiplier-tube size, and positioning, as well as the design or linearity of the electronic circuits. For the three classes of cameras examined (Nuclear Chicago Pho Gamma III, Nuclear Chicago High Performance, and Picker 2 or 2C) the counting-rate performance was less than one half that which can be expected with better electronics. Resolution without barrel distortion was maintained best by the Picker cameras. However, the Nuclear Chicago High Performance camera gives more than twice the count rate for the same amount of isotope.

Resolution is a sensitive function of distance from the collimator, and for cardiac clinical studies is practically independent of absorber between the organ and the scintillation camera. This observation is limited to scintillation cameras with a 20–30% window centered on the photopeak. Resolution of gamma imaging systems deteriorates very rapidly as absorber thickness increases. For cardiac flow studies one can expect $\frac{1}{4}$-in. resolution and no serious distortion of the center field of view as count rates increase to over 80,000 cps. This corresponds to injection of 40 mCi of isotope and a gross event rate on the crystal of 300,000 cps.

The response of a NaI(Tl) crystal has been shown in theory and in practice to perform at over 500,000 gross events/second. Not only can the camera respond to far more isotope than conventionally used, but the electronics of commercial cameras can be improved to give two or more times the statistics for a given injected dose. There should be no serious loss in resolution during high-speed cardiac flow studies using two to four times the conventional millicurie amount of isotope. The statistics needed for quantitation of flow patterns and anatomical distributions can be obtained with the scintillation camera.

HIGH-COUNT-RATE CARDIAC STUDIES WITH A MULTICRYSTAL GAMMA-CAMERA

The time required for accurate electronic positioning of observed events causes the relative inefficiency of single-crystal gamma-cameras for high count rate applications (6). An alternate approach to instrument design which promises to increase greatly the counting efficiency of stationary imaging devices utilizes a matrix of individual crystals which remain constant in position, thereby eliminating the electronic time required for spatial orientation of observed counts. A new multicrystal gamma-camera has been demonstrated to achieve greater than 200,000 cps (7). The high count rate possible with this instrument appears to greatly enhance the hemodynamic information available from radionuclide cardiac studies in patients.

Instrument Design

The Baird-Atomic System Seventy Scanning Gamma Camera, which is basically a matrix of scintillation crystals interfaced directly into a small computer, incorporates a number of features used in the older Digital Autoflouroscope designed by Bender (8). Collimation for the detector is provided by lead slabs which can be arranged in 1.0-, 1.5-, and 2.5-in. thicknesses. Centered over each of 294 crystals in a 14 × 21 matrix are tapered holes with square cross section of 0.250-in. sides at the crystal surface, 0.213-in. sides at 1.0 in., 0.194-in. sides at 1.5 in., and 0.157-in. sides at 2.5 in distance. The 14 × 21 array of 1 cm² NaI(T) scintillation crystals with a 4-cm depth are connected by Lucite light pipes to 35 photomultiplier tubes (one for each column and row) for event positioning. Pulses from 35 preamplifiers and amplifiers serving these photomultiplier tubes pass to 35 lower-level discriminators which eliminate events observed simultaneously in adjacent crystals. This process requires only 100 nsec in any of the 35 lower-level discriminators and insures that only gamma interactions which occur within a single crystal are recognized as valid. Recognition of an invalid event by a lower-level discriminator immediately resets the single channel analyzer to minimize dead-time from anticoincidence processing. Energy analysis of amplifier pulses is performed with an event-resolving time of 0.5 μsec.

Valid events are stored in a random-access, solid-state buffer memory. Information in binary form representing data from the entire detector passes into the memory of an 8000 16-bit word Nova computer as rapidly as every 50 msec. A 100,000 word magnetic drum provides immediate data storage. Capacity of the Nova computer may be expanded to 32,000 words, and drum memory with 1,500,000 words is available. Optional data storage devices include cassette tape and IBM compatible magnetic tape. A variety of peripheral devices process and display the data. Computer programs for data accumulation, correction, manipulation, and display are retained on a portion of drum memory and are executed by hardwired command functions.

Counting-Rate Characteristics

The long electronic dead-time for event processing represents the greatest limitation of camera-type detectors for dynamic radionuclide studies. Instrument dead-time is a complex parameter which includes the resolving time of all individual components of the total system. A more practical term describing instrument performance is event-resolution time which sums all component parameters and is the minimum time required between two processed events. The efficiency of most gamma-cameras, as defined at low count rates, declines linearly with increasing counting rates. Extrapolation of the decrease in detection efficiency with increasing radioactivity defines a theoretic maximum counting rate of the instrument. The reciprocal of the maximum counting rate defines the processing time required between two events or the event resolution time. At counting rates below the theoretic maximum, single-crystal gamma-cameras demonstrate spatial distortion which results primarily from coincidence-summing of two 74-keV lead K shell x-rays. At the point of detector saturation, gamma-cameras demonstrate a decreasing count rate with increasing radioactivity. The counting rate at which instrument saturation occurs determines the maximum usable counting rate which is a second performance characteristic important in dynamic radionuclide studies.

A simple experiment was designed to evaluate the count-rate characteristics of the System 70 matrix camera and the "high performance" single crystal Anger camera. A small radioactive source was introduced into the detector field and radiation was gradually increased over a distant region of the detector. Individual crystal counts were examined as a function of the total detector counts. At counting rates less than 180,000 cps, the matrix camera responds as if it would reach a maximum of about 450,000 cps which defines an event-resolution time of $2\frac{1}{4}$ μsec (Fig. 159). However, the detector actually saturates at 230,000 cps and provides usable data only up to about 200,000 cps. The experiment repeated on the Anger camera demonstrated a dead-time of 12 μsec, and saturation at 50,000 cps (Fig. 160).

The counting efficiencies of the matrix camera with three different collimators and the Anger camera with the high-resolution collimator were examined by adding increasing amounts of ^{99m}Tc pertechnetate to a phantom within the field of detector view (Fig. 161). The matrix camera records counts in binary form within computer memory, and retrieval of digital information causes no further data loss. Counts observed on the Anger camera represent only scaler readings, and further count loss or data degradation by electronic noise would occur in organizing and storing data in binary form. The matrix camera with the 1.0-in. collimator reached count saturation with about 15 mCi ^{99m}Tc and lost counts with further increases in radioactivity due to anticoincidence processing time. The 1.5-in. collimator allowed about 150,000 cps with 15 mCi and reached nearly 200,000 cps with 30 mCi of ^{99m}Tc. The 2.5-in. collimator passed 100,000 cps with 40 mCi of ^{99m}Tc. The 1.0, 1.5, and 2.5 in. System 70 collimators with the matrix camera demonstrated efficiency ratios of 9.8:3.9:1.0. The Nuclear Chicago Pho/Gamma-HP provided only about 40,000 cps with 40 mCi of ^{99m}Tc and a high-resolution collimator of 2-in. depth. Use of the high-sensitivity (1-in.) collimator with the Anger camera would provide a threefold greater count rate at low doses, but would not alter the maximum useful counting rate of less than 50,000 cps.

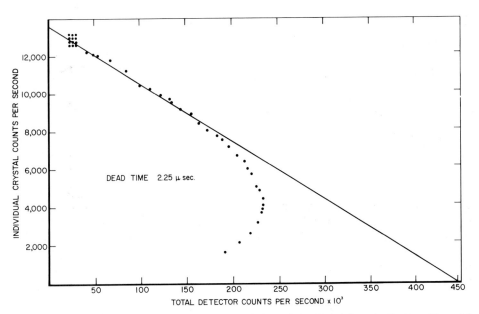

Fig. 159. Data describing count-rate characteristics of the Baird-Atomic System 70 matrix camera. Counts from a constant radiation source at a single crystal location decline with increasing total detector count rate. These data provide both the theoretic maximum counting rate and the counting rate causing detector saturation.

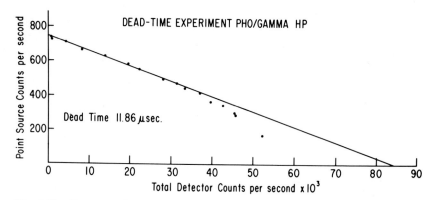

Fig. 160. Data describing count-rate characteristics of the Nuclear Chicago Pho/Gamma HP Anger camera.

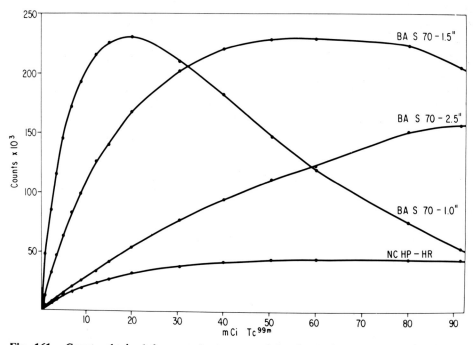

Fig. 161. Counts obtained from a phantom containing increasing concentrations of ^{99m}Tc-pertechnetate compare the counting efficiency of the Baird-Atomic System (BA-@70) with 1.0, 1.5, and 2.5 in. collimators and the Nuclear Chicago Pho/Gamma HP (NC-HP) with the low-energy high-resolution (HR) collimator.

REFERENCES

1. J. P. Kriss, W. A. Bonner, and E. C. Levinthal: Variable Time-Lapse Videoscintiscope: A Modification of the Scintillation Camera Designed for Rapid Flow Studies, *J. Nucl. Med.* **10:** 249, 1969.
2. P. M. Weber, L. V. Dos Remedios, and I. A. Jasko: Quantitative Radioisotopic Angiocardiography, *J. Nucl. Med.* **13:** 815–822, 1972.
3. Y. Ishii, and W. J. MacIntyre: Measurement of Heart Chamber Volumes by Analysis of Dilution Curves Simultaneously Recorded by Scintillation Camera, *Circulation* **44:** 37–46, 1971.
4. R. H. Jones, D. C. Sabiston, B. B. Bates, J. J. Morris, P. A. W. Anderson, J. K. Goodrich: Quantitative Radionuclide Angiocardiography for Determination of Chamber to Chamber Cardiac Transfer Time, *Am. J. Cardiol.* **30:** 855–864, 1972.
5. J. P. Kriss, L. P. Enright, W. G. Hayden, et al.: Radioisotopic Angiocardiography—Wide Scope of Applicability in Diagnosis and Evaluation of Therapy in Disease of the Heart and Great Vessels, *Circulation* **43:** 792–807, 1971.
6. R. H. Jones, et al: Basic Considerations in Computer Use for Dynamic Quantitative Radionuclide Studies, Proceedings of 2nd Symposium on Sharing of Computer Programs and Technology in Nuclear Medicine, Oak Ridge, Tennessee, 1972, p. 133, Conf-720430.
7. R. H. Jones, R. P. Grenier, and D. C. Sabiston, Jr.: Description of A New High Count-Rate Gamma-Camera System, *Medical Radioisotopes Scintigraphy 1972*, **1:** 299–312, 1973.
8. M. A. Bender, and M. Blau.: The Autofluoroscope, *Nucleonics* **21:** 52, 1963.

PRINCIPAL EDITOR: RAO CHERVU

9. Radiopharmaceuticals in Radiocardiography

The growth of the methods of radiocardiography for the functional assessment of the heart with radionuclides has been phenomenal during the last few years (1–12). This is mainly due to parallel developments in the field of radiopharmaceuticals and in high-resolution imaging devices. The installation at a few centers of compact medical cyclotrons with large beam currents and a diversity of charged particles that can be accelerated has widened the scope of utilization of short-lived and ultrashort-lived radionuclides produced through low- and medium-energy nuclear reactions (13, 13a). The availability of these radionuclides has added a new dimension to the application of radiopharmaceuticals by minimizing the radiation dose to the patient. With sufficiently rapidly decaying nuclides, serial examinations can be conducted within minutes or hours; judicious choice of radionuclides permits use of serially higher gamma energies for studies done minutes apart or use of inhaled materials to reach the left heart directly.

Quantitation of coronary blood flow with new radioisotope techniques has permitted new ways of evaluation of blood supply to the myocardium. Delineation of regional myocardial perfusion and function has been previewed by imaginative use of selected radiopharmaceuticals. A new methodology for evaluation of disorders of the heart and great vessels has evolved with these noninvasive radioisotopic methods; developments in radiopharmacy have played a major and continuing role in this development.

CRITERIA FOR SELECTION OF RADIOPHARMACEUTICALS

Physical Characteristics

Radionuclides for clinical application must satisfy the following criteria. The half-life of any radionuclide used for the assessment of a certain physiological function should be such that the average life of the nuclide ($1.433 \times T_{1/2}$) should approach the length of the time required to measure the function (14). Since the duration of a typical radiocardiographic evaluation is generally of the order of 10 minutes, a short-lived radionuclide is indicated. In order to minimize radiation dose to the patient, particularly in repeat studies, the effective half-life of the radionuclide must be short. For example, ^{85}Kr and ^{133}Xe have relatively long physical half-lives of 10.76 years and 5.3 days, respectively (15), but very short biological residences and hence very short effective half-lives. The radionuclide must have minimal or absent emission of particulate radiation; thus nuclides decaying by electron capture or isomeric transition are preferred. In some cases, positron emitters have certain advantages since the annihilation radiation of 511 keV is suitable for coincidence counting which lends itself to spatial resolution not otherwise attainable, but most crystals in current use have low efficiency for this energy, a fact which may be compensated by use of very large doses of very short-lived elements.

The energy of the gamma rays emitted should be high enough for ease in spectroscopy and low enough (<300 keV) for high photopeak counting efficiency in the $\frac{1}{2}$-in. crystal of the gamma-camera. Very-low-energy gamma radiation (<100 keV) leads to excessive narrow-angle scatter in both collimator and crystal, and higher probability of tissue absorption within the patient, contributing to patient radiation but not to diagnostic information. These considerations indicate an optimal gamma energy range of about 200 keV. The short-lived nature of the optimal nuclide (<20 minutes) dictates its production in a cyclotron or a reactor close to the site of clinical application (impractical at many centers) or, alternatively, one of a number of generator systems which have reasonably short-lived and chemically versatile daughter products (^{99m}Tc, ^{113m}In, etc.) with favorable nuclear decay characteristics.

Physiological Considerations

The radioactive tracers used in radiocardiography studies of regional blood flow are broadly classified under two categories:

a. Large non-diffusible molecules (^{99m}Tc-albumin and red cells, and ^{113m}In-transferrin) which remain within the vascular system and have a volume of distribution confined to the circulating blood.

b. Diffusible substances ($^{43}K^+$, $^{99m}TcO_4^-$, ^{85}Kr) which pass through the capillary walls and equilibrate with the extravascular or intracellular spaces. The choice of agent is dependent upon the objective of the particular study.

Practical Considerations

For blood-pool measurements which are normally completed within a time interval of 10 to 30 minutes from the time of injection, the half-life of the nuclide must be sufficiently long and, of course, the tracer must be incorporated in a nondiffusible molecule. It may be desirable to repeat the tracer study within a few hours or days; background activity from the previous study should be negligible. For many dynamic-flow studies, diffusible ions like ^{99m}Tc-pertechnetate are quite suitable and the shorter biological half-life makes them preferable both in reducing radiation dose and in reducing background for serial studies. Radioactive gases which are extensively used in shunt studies have been considered from the viewpoint of their availability, convenience, and safety.

The quanta of radioactivity that are administered in flow studies have to be available in high specific concentrations (5–10 mCi/ml) such that a bolus injection is possible. The bolus injection allows the tracer to remain well defined during its entry into the first mixing chamber of the heart, permitting temporal separation of down-stream chambers as defined in Chapters 2–4. These considerations define the criteria for choice of radionuclide for radioisotopic angiocardiography, coronary angiography, and myocardial blood flow studies. Pertinent physical data for the various radionuclides proposed for these studies are given in Table 8.

AGENTS FOR ANGIOCARDIOGRAPHY

^{99m}Tc-Pertechnetate

^{99m}Tc-pertechnetate is available from sterile ^{99}Mo-^{99m}Tc generators (16) in high specific concentration of 20 mCi/ml. The fractional elution technique shown in Figure 162 indicates how a smaller volume of peak activity can be used to achieve greater concentration than the average eluate, if necessary. Instant technetium is available at any desired specific concentration (50 mCi/ml or more), commercially delivered on a daily basis.

^{99m}Tc Human Serum Albumin

^{131}I human serum albumin has been used for many years as a radioactive indicator for measurement of cardiac output by external counting, but has a long effective half-life (6 days) and abundant beta decay resulting in a high patient radiation dose. By contrast, ^{99m}Tc albumin is well suited to these studies. For preparation, ^{99m}Tc-pertechnetate is reduced to a lower valence state of ^{99m}Tc (IV or V) through the use of various agents (17–22) and complexed with serum albumin, or it may be directly electrolytically complexed with human serum albumin (23). The electrolytic complexation process of ^{99m}Tc-albumin for rapid closed-system production (24) is a convenient adaptation of the original process which achieves 88 to 95% albumin binding which remains stable for 30 to 60 minutes after administration.

The method of preparation of ^{99m}Tc-HSA using Sn(II) ion as a reductant involves the following steps (22). 1 ml of 250 mg/ml of HSA is added to 0.5 ml of 2 mg/

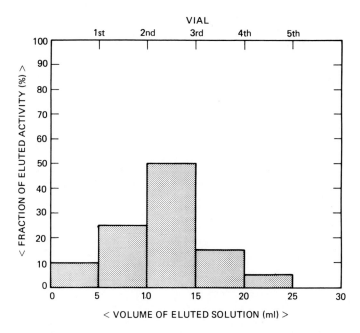

Fig. 162. Fractional elution of ^{99m}Tc from ^{99}Mo–^{99m}Tc generator. Position and height of the peak activity varies according to the generator.

ml of freshly prepared $SnCl_2 \cdot 2H_2O$ solution, followed by addition of 0.1 to 2 ml of $^{99m}TcO_4-$, and the solution is mixed for 1 minute. The pH is raised to 6 with 1 N NaOH. Other variations of the Sn–^{99m}Tc-HSA procedure have been reported for the preparation of high-specific-activity ^{99m}Tc-HSA using smaller amounts of HSA (25 mg) and 30 minutes of stirring the reaction mixture after addition of pertechnetate (25). The binding efficiency using this procedure is greater than 95%. Sterilization of the final preparation is achieved by terminal millipore filtration through a 0.22-μ filter.

The degree and nature of Tc binding with albumin varies with the method of preparation, and the material may have more than one component (24, 26, 27). Simultaneous blood-volume determinations using ^{99m}Tc-albumin and iodinated human serum albumin indicate close agreement (7, 26, 28).

^{99m}Tc-Labeled Erythrocytes

Labeling of red blood cells with ^{99m}Tc has been reported by several workers (29–37). The problem involves two steps: (1) introducing the technetium into the cells and (2) firmly binding the same within the cells. Several variations of these two steps are proposed with somewhat conflicting results.

Labeling efficiency of up to 97% is reported (36) with the following method: 8 ml of blood from the patient are withdrawn into a sterile capped tube containing 2 ml of a 3.8% sodium citrate solution, and cells and plasma are separated by centri-

Table 8. Radionuclides of Interest in Radiocardiography and Relevant Properties [a]

Nuclide	Typical Production Method	Half Life	Decay Mode	Major Photon Energies Kev	Fractional decay [b]	Comments
^{11}C	^{10}B$(d,n)^{11}$C	20.3 minutes	β^+	511	(200%)	Accelerator produced on-site; labeled CO and CO_2 gases in millicurie quantities; rapid chemistry required to form different compounds.
^{13}N	^{12}C$(d,n)^{13}$N	9.96 minutes	β^+	511	(200%)	Accelerator produced on site; molecular N_2 gas or in solution or as labeled ammonium ion.
^{15}O	^{14}N$(d,n)^{15}$O	2.03 minutes	β^+	511	(200%)	Accelerator produced on-site; available as $^{15}O_2$, $C^{15}O$, $C^{15}O_2$ gases.
^{42}K	^{41}K$(n,\gamma)^{42}$K	12.4 hours	β^-	1524	(18%)	High energy and poor detection efficiency; very large radiation dose to the patient.
^{43}K	^{40}Ar$(\alpha,p)^{43}$K	22.4 hours	β^-	373 390 590 619	(85%) (18%) (13%) (81%)	Accelerator produced; may find wide application for myocardial imaging and function studies; ^{42}K contamination problem for clinical application. Commercially available soon.
^{81m}Kr	^{81}Rb generator milked	13 seconds	IT	190	(65%)	Millicurie activities available; short halflife of generator a disadvantage.
^{85m}Kr	^{84}Kr$(n,\gamma)^{85}$Kr	4.4 hours	β^-, IT	150 305	(74%) (13%)	Inert gas solution, ideal photon energies; suitable for circulation and blood flow studies, short half life a serious disadvantage.
^{85}Kr	Fission gas	10.76 years	β^-	514	(0.41%)	Inert gas solution; poor external detection by gamma counting; available commercially.

Isotope	Production	Half-life	Decay mode	γ energy	Fractional decay	Comments
^{81}Rb	^{79}Br(α, 2n)^{81}Rb	4.7 hours	E.C., β^+	511	(26%)	Millicurie activities produced. Soon to be made available commercially.
				190	(65%)	
^{84}Rb	^{81}Br(α, n)^{84}Rb	33 days	E.C., β^+	511	(42%)	Long half life; high radiation dose.
			β^-	880	(74%)	
^{86}Rb	^{85}Rb(n, γ)^{86}Rb	18.7 days	β^-	1080	(8.8%)	Long half life; high radiation dose.
^{99m}Tc	^{99}Mo generator produced	6 hours	IT	140	(90%)	Versatile chemical characteristics, available in high specific concentration commercially.
^{113m}In	^{113}Sn generator produced	100 minutes	IT	393	(64%)	Excellent chemical characteristics, commercially available; reasonable specific concentration in milkings.
^{123}I	^{122}Te(α, 3n)^{123}Xe ^{123}Xe → ^{123}I	13.3 hours	E.C.	159	(83%)	Not available commercially at reasonable cost; very good chemical behavior.
^{127}Xe	^{127}I(p, n)^{127}Xe	36.4 days	E.C.	172	(22%)	Commercially not available; good imaging characteristics; low radiation dose.
				203	(65%)	
				375	(20%)	
^{133}Xe	Fission gas	5.3 days	β^-	81	(37%)	Widely used in several studies; available commercially.
^{129}Cs	^{127}I(α, 2n)^{129}Cs	32.1 hours	E.C.	375	(48%)	Cyclotron produced; may find application for myocardial studies; soon available commercially.
				416	(25%)	
^{131}Cs	^{130}Ba(n, γ)^{131}Ba ^{131}Ba → ^{131}Cs	9.7 days	E.C., β^-	Xe X-rays		Very low energy; not a particularly useful agent for myocardial imaging; production method cumbersome.
^{137m}Ba	^{137}Cs generator produced	2.5 minutes	IT	662	(89%)	High energy and poor detection efficiency; shortlived and hence useful; generator not available commercially.
^{201}Tl	^{203}Tl(p, 3n)^{201}Pb ^{201}Pb → ^{201}Tl	74 hours	EC	70 (Hg x-rays)		Low energy photon emission not ideal for detection; commercially unavailable.

[a] Nuclear properties listed from reference 15.
[b] Fractional decay, the fraction of decays resulting in a given γ energy, is referred to as "intensity" in some handbooks.

fuging for 10 minutes at 1500 rpm. The red cells are washed once with saline and incubated with $SnCl_2$ (0.5 μ Sn^{2+} in about 0.5 ml of a 0.9% NaCl solution) in a water bath at 37°C for 5 minutes. ^{99m}Tc-pertechnetate (<2 ml) is now added and the mixture incubated for 5 minutes at 37°C. The labeled cells are washed once with saline and resuspended in saline or the patient's own plasma prior to injection.

The red blood cells should not be allowed to stand packed for prolonged time intervals, to avoid agglutination. In vitro and in vivo stability of the Tc-labeled red blood cells is found to be very good over the course of 30 minutes, during which time simultaneous determination of red cell volume in man with ^{51}Cr-labeled erythrocytes and ^{99m}Tc-labeled erythrocytes has given nearly identical results (38). Thereafter, the technetium label distributes progressively into a slightly larger space compared to the ^{51}Cr label. In the case of generator-eluted $^{99m}TcO_4^-$, the presence of ionic aluminum(III) contamination in the eluate might cause agglutination of erythrocytes (39).

Though the ^{99m}Tc-labeled red blood cells are a useful preparation for blood-volume determinations, the specific concentration achieved is not sufficiently high for dynamic-flow studies. More efficient methods of tagging red cells with higher activities of pertechnetate need to be developed.

Indium-113m Transferrin

The ^{113}Sn-^{113m}In generator system was introduced by Stern et al. (40) into clinical medicine in 1966. ^{113m}In is eluted as indium chloride from the generator system with dilute HCl (0.05 N); this nuclide may be incorporated into several radiopharmaceuticals for organ imaging (40–44). The physical properties of ^{113m}In are given in Table 7. Large-size generators are available commercially and ^{113m}In concentrations of about 10 mCi/ml are obtained easily, which is quite adequate for radiocardiographic studies. However, when the generator yields of ^{113m}In are of low specific concentration, it may not be practical to increase the specific concentration by rapid evaporation of the eluate in the absence of the carrier.

^{113m}In chloride solution is bound specifically and tightly to serum transferrin both in vivo and in vitro (45, 46). A plasma sample is labeled with ^{113m}In by mixing a quantity of ^{113m}In chloride in acidic gelatin solution; or a small quantity of ^{113m}In eluted directly from the generator may be injected, which will undergo virtually instantaneous binding to transferrin (45, 46) as long as the patient's transferrin is not already saturated with iron. The use of ^{113m}In transferrin for measurement of cardiac output is very practical from the standpoint of ease of labeling and low radiation dose to the patient (47). The determination of the plasma volume can be carried out simultaneously, provided sampling is done during the first 10 minutes (28, 46). During this period, it is assumed that the administered activity does not leave the plasma space. The short physical half-life of ^{113m}In (100 minutes) requires that the dose be used promptly once eluted from the generator. The values obtained for plasma volume with ^{113m}In-transferrin average 5% higher than those obtained with iodinated albumin. Owing to the hydrolytic behavior of ^{113m}In in very dilute acid solution, and consequent loss due to absorption on glass or plastic, dilutions, when necessary, are performed with 0.01 N HCl solutions.

^{113m}In-DTPA

^{113m}In-DTPA complex has also been used for cardiac function study (48). The complex is prepared easily within 10 to 15 minutes by the procedure of O'Mara et al. (49). ^{113m}In-DTPA is rapidly cleared from the blood stream by the kidneys, reducing the background for repeat studies.

The activity eluted from the generator is placed in a centrifuge tube and 1 ml of DTPA pentasodium salt (8 mg/ml) is added while stirring. The pH is adjusted to 7.0 with 1 N NaOH and the mixture stirred for an additional 2 minutes. The preparation is sterilized by millipore filtration (0.45-μ pore size).

Iodine-123

Among the iodine isotopes, ^{123}I fulfills the criteria of the ideal gamma isotope for *in vivo* diagnostic procedures more closely than any other of its radionuclides (50). The physical properties of this nuclide are given in Table 7. The relatively short half-life of 13.3 hours and the emission of a 159-keV photon (83% abundance) by this nuclide are both attractive features, in terms of tissue penetration and detection by high-resolution counting equipment, and in terms of low radiation dose and possibility of repeat studies.

Iodine-123 is produced through a wide variety of nuclear reactions (51–58) with varying degree of contamination from other iodine isotopes, and with varying yields.

^{123}I is produced through the decay of 2.1-hour ^{123}Xe, which in turn is formed either by ^{122}Te (α, 3n)^{123}Xe or ^{127}I(p, 5n)^{123}Xe reactions, available in a high-purity, nearly carrier-free state. The preparation of a number of iodine-labeled radiopharmaceuticals formerly labeled with ^{131}I are being contemplated with increasing availability of ^{123}I. Various methods of iodination of albumin and other proteins are reported of which the most frequently used synthetic methods are iodination by KI_3 (59, 60), use of iodine monochloride (61–68), electrolytic iodination (69), and chloramine-T method (70, 71) or enzymatic iodination (72). The iodine monochloride method of labeling human serum albumin and other radiopharmaceuticals has been well adapted to the short-lived ^{123}I radionuclide (73). Affinity of the tyrosine molecule for iodine results in firm and stable albumin labeling, with over 97% binding achieved routinely, a significant improvement over Tc-albumin complexing. High overall radiochemical yield and high specific activities (700 μCi/mg) starting with 1 mCi of ^{123}I have been obtained using the ICl methods (73).

Although ^{123}I can be cyclotron-produced directly, the high-energy contaminant ^{124}I presents a substantial problem except when formed through the indirect reaction through ^{123}Xe decay (74, 75). The routine supply of high-purity ^{123}I preparations, at a reasonable cost, in the millicurie quantities needed for radiocardiography, awaits the large-scale production consequent to planned, but as yet nonoperating, facilities (June 1974). Curie quantities of high purity ^{123}I are expected to be continuously produced at the Brookhaven Linac Isotope Producer (BLIP) facility (76).

Barium-137m

Barium-137m has excellent dosimetric characteristics, but its chemical properties in terms of its binding to the vascular pool and its high photon energy (662 keV) are far from ideal. Barium and other alkaline earth metals are much less versatile in their coordination affinities than the transition metals and hence are not strongly bound to serum albumin (77). The nuclide may find application in step-up studies, when only first transit vascular imaging is required and a higher energy photon is desirable in order to decrease a high patient background, due to prior ^{99m}Tc and ^{113m}In doses. The photo peak efficiency of $\frac{1}{2}$-in. NaI crystal for the 662-keV photon of ^{137m}Ba is only 10%. However, several workers have used this nuclide to obtain quantitative data on various parameters of cardiac function (78–81).

Cesium-137 ($T_{1/2} = 30$ years) decays to an isomeric state of ^{137m}Ba ($T_{1/2} = 2.6$ minutes) which emits a gamma ray of 662 keV of 89% abundance (15). High specific concentrations of ^{137m}Ba are achieved with certain generator designs (82–85), including holdup resin columns to insure minimum contamination from the long-lived ^{137}Cs parent nuclide. Possibility of leakage of long lived ^{137}Cs from the column due to the long-term radiation effect of repeated elutions on this breakthrough is a factor that has not been fully established for wide application of this radionuclide in various rapid dynamic clinical studies.

Radioactive Gases

A large number of radioactive isotopes of the noble gases krypton and xenon are formed in neutron fission or through nuclear reactions on uranium. The two radionuclides ^{85}Kr and ^{133}Xe have found extensive application in pulmonary physiology and radiocardiography studies (86). At centers where cyclotron facilities exist, several other radionuclides available as the gases (O, CO_2, N_2, and O_2) which are short lived, ^{11}CO (20.3 minutes), ^{13}N (9.96 minutes), and ^{15}O (2.03 minutes) can be produced. Use of these short-lived radionuclidic gases is limited to centers close to the site of production.

Carbon-11 Labeled Gases

Carbon-11 has a half-life of 20.3 minutes and decays by emission of 0.96 MeV β^+ particles (15). The positron annihilation gammas are emitted with 200% abundance. The very high affinity of CO for hemoglobin and the rapid diffusion of CO across the alveolar membrane, combined with the need for a tracer which can be "injected" into the left atrium, have stimulated development of ^{11}C as a radiocardiographic pharmaceutical. The problems of delivery, of administration in a time-pulsed manner so that only one breath would make input, and of collimation sufficient to separate the heart from the lung and airways have not yet been dealt with quantitatively. However, the promise is substantial to those in the vicinity of a cyclotron, in almost any of which ^{11}C is a highly abundant product.

Carbon-11 is produced through several nuclear reactions, ^{10}B (d, n) ^{11}C, ^{11}B (d, 2n) ^{11}C, ^{11}B (p, n) ^{11}C, ^{14}N (d, α n) ^{11}C, ^{14}N (p, α) ^{11}C, in practicable quantities in a cyclotron (87–92). Details of production of ^{11}CO and $^{11}CO_2$ for routine clinical use and storage devices are given by Clark and Buckingham (92). In the routine production of ^{11}CO, ^{11}C hot atoms generated in the nuclear reaction are swept away from the target chamber using 1% CO in He and are freed from CO_2 in a reducing furnace. If H_2 is used as a sweep gas, very little $^{11}CO_2$ is formed, obviating the need for a reducing furnace as well as having the advantage of a very low $^{13}N_2$-contaminant level. The presence of CO carrier lowers the specific activity of the target effluent gas and causes inefficient red-cell labeling. A ^{11}CO storage system has been described in which the labeled gas ($>$150-mCi charge) may be trapped on either activated charcoal at $-196°C$ or on a molecular sieve at $-85°C$ from which the gas can be released at the user's site several miles away from the cyclotron. $^{11}CO_2$ may be produced continuously without the use of the reducing furnace, using either a sweep gas of pure He or 1% CO_2 in He, and stored conveniently. Typical radioactive concentrations released from the storage system are approximately 3.5 and 10 mCi/ml for ^{11}CO and $^{11}CO_2$, respectively (92).

The method for labeling of red blood cells (93) with ^{11}CO is as follows: 5 ml of venous blood is withdrawn into a 50-ml syringe containing 1.5 ml of acid citrate dextrose solution as the anticoagulant. 45 ml of ^{11}CO plus helium carrier gas mixture is withdrawn into the syringe which is then rotated at 10 rpm for 10 minutes after which excess gas is expelled. The amount of the ^{11}CO contained in the plasma is less than 0.01% and hence it is not necessary to wash the plasma from red cells after labeling. Specific concentration of up to 75 μCi/ml of blood are achieved using this procedure. The specific concentration is not sufficiently high from the view point of application to radiocardiography and blood-pool studies, but it should be possible to attain higher specific concentrations. Measurement of blood volume using red cells labeled with ^{11}CO have been reported (93, 94).

Krypton Radionuclides

Krypton-85, a beta-emitting isotope of half-life 10.3 years with low-intensity gamma radiation of 0.514 MeV (0.41%) is available commercially in gaseous form or as a saline solution in multidose vials for intravenous administration. The solubility of ^{85}Kr (95–97) in blood at 37°C is very low (0.058 ml/ml blood) and, as such, over 95% of the ^{85}Kr administered intravenously is exhaled on its first passage through the lungs, consequently minimizing the recirculation component. The solubility of ^{85}Kr in fat (98) is given as 0.51 ml/g at 37°C. External counting of the activity of ^{85}Kr is not feasible owing to the low abundance of the 514-keV gamma radiation. Blood samples and aliquots of standards are transferred to air-tight chambers fitted over the end window beta counters for measurement of beta activities. In this respect, ^{85m}Kr ($T_{1/2} =$ 4.4 hour, gamma energy 150 keV) might prove useful for wider application, though the short half-life is a serious limitation.

Krypton-81m, a short-lived radionuclide ($T_{1/2}$ = 13 seconds) is generator produced from ^{81}Rb and decays by isomeric transition with the emission of 190-keV gamma rays (65% abundance) to long-lived ^{81}Kr (15). Krypton-81m has been employed for multiple radiocardiography studies with low radiation exposure (99–101).

Radionuclides of Xenon

Xenon-133 is a reactor-produced nuclide and decays with a half-life of 5.3 days, emitting gamma rays of energy 80 keV (37%) and x-rays of 31 keV (15). Xenon gas is soluble to a limited extent in plasma and blood at 37°C (α = 0.10 and 0.18, respectively) and more so in fatty tissues (α = 1.83) (96–98, 102, 103). The gas is removed from the circulation on its first passage through the lungs to a large extent (95%) after intravenous administration.

Xenon-133, though readily available, is not ideal with regard to its radiation characteristics; the narrow angle scatter of the primary photons of 80 and 30 keV and the poor depth response of the low-energy radiation are undesirable. Xenon-127 ($T_{1/2}$ 36.4 days; and E.C. mode of decay) emits principally 172 keV (22%), 203 keV (65%), and 375 keV (20%) photons (15) and would seem to offer better radiation characteristics for gamma-camera imaging and lower radiation dose to the patient (104, 105). Xenon-127 is produced through the reactions ^{127}I(p, n),^{127}I ($d, 2n$) or by ^{126}Xe(n, γ) reactions (15). At present this nuclide is not available commercially. It is proposed to produce it in multicurie quantities at the BLIP facility through the ^{133}Cs($p, 2p5n$)^{127}Xe reaction (76).

The recovery and reconcentration of the xenon gas for storage or for reuse is possible with the development of suitable high-efficiency adsorbent systems or cryogenic traps. Systems based on these principles are reported by Liuzzi et al. (106) and Corrigan et al. (107, 108).

Xenon-133 is commercially available in high specific activity as a gas in glass ampules or in saline solutions in multiple-dose vials at a high specific concentration. Several techniques have been described for handling curie quantities of ^{133}Xe and for dispensing individual doses as gas and in saline for administration to the patient (109–115). These provide a high degree of flexibility regarding the specific concentration desired for various clinical purposes at a relatively low cost. Being a relatively poorly soluble gas in saline (α = 0.078) (97), many precautions for storage and dispensing of saline solutions of ^{133}Xe from multidose vials are necessary (116–118). Xenon leakage during storage of saline solution could be considerably reduced by inverting and chilling the bottle to prevent leakage from the rubber stoppered multi-injection bottle. Even a small amount of air introduced into the storage bottle will cause a decrease in the amount of dissolved ^{133}Xe radioactivity. Syringes and needles used in preparation should be rinsed with saline to remove all trapped air, leaving saline in syringe hub and needle prior to drawing a dose from the stock bottle. In plastic syringes, doses should be drawn just prior to injection wherever possible to avoid leakage of xenon through the plastic, which occurs with a half time in the order of 120 minutes (117).

Nitrogen-13

The longest-lived radionuclide among the nitrogen isotopes is ^{13}N, ($T_{1/2}$ = 9.96 minutes) which decays by positron emission resulting in annihilation radiation (200%). This is only useful for application in a clinical facility in the vicinity of a cyclotron. Nitrogen-13 is produced in a cyclotron through the ^{12}C(d, n) reaction (88, 89, 119–121). Details of the method of producing sterile solutions labeled with ^{13}N$_2$ are given by Buckingham and Clark (121). Specific concentrations of up to 300 µCi/ml are achieved using a 14-MeV deuteron beam current of 60 µA. The solubility of ^{13}N$_2$ is one fourteenth that of ^{133}Xe in blood (α = 0.013) and thus nitrogen is cleared from the circulation very rapidly. Furthermore, coincidence counting with its excellent spatial resolution can be employed (122). Thus the application of this nuclide could be envisaged at those centers having a cyclotron facility on site. Nitrogen-13 labeled ammonia has been suggested as a valuable agent for myocardial imaging (see Chapter 6).

Oxygen-15 Labeled Gases

Oxygen-15, the longest-lived radioisotope of oxygen, has a half-life of 2.1 minutes and decays by positron emission (100%). The use of this short-lived isotope in cardiac dynamic studies and pulmonary physiology has so far been exceedingly important to research in cardiopulmonary physiology, although limited to a very few institutions having cyclotron facilities in direct conjunction with a patient study area. In the foreseeable future, this technique can only be considered for research, as opposed to clinical use. The uptake, distribution, and clearance rate of the ^{15}O$_2$, C^{15}O, and C^{15}O$_2$ gases in the body depends upon the method and site of introduction of the activity in the body, on the solubility of the gases introduced in blood and tissues, and on the perfusion characteristics of the various organs. The application of these gases has been chiefly in pulmonary physiology including measurement of regional blood flow and detection of cardiac shunts (110, 123–126); ^{15}O-carbon dioxide, being very soluble, rapidly passes into the pulmonary capillary blood when inhaled, and this inhalation technique has been employed for obtaining inhalation radiocardiograms by Jones et al. (127). The technique has a distinctive advantage for left-heart studies since there is no activity in the right heart. The short half-life of 2.1 minutes enables millicurie amounts of ^{15}O to be administered, thereby providing satisfactory statistics and allowing measurements at frequent intervals.

The continuous production of ^{15}O is achieved in a cyclotron by bombarding nitrogen gas with low-energy deuterons through the reaction ^{14}N (d, n) ^{15}O (13, 88, 89, 119). Oxygen-15 labeled carbon dioxide is prepared by passing the oxygen through activated charcoal at 500°C and subsequently passing the labeled carbon dioxide over cupric oxide to oxidize any carbon monoxide present in the gas. Oxygen-15 labeled carbon monoxide is produced by passing the gas from the target gas chamber over activated charcoal at 850°C and removing the traces of carbon dioxide by absorption by sodalime.

AGENTS FOR STUDY OF CORONARY CIRCULATION AND MYOCARDIAL IMAGING

A variety of radioisotopic techniques for myocardial scanning and blood flow are at present available which include (*a*) the use of ions or agents that are incorporated into the myocardial cell and (*b*) use of radiopharmaceuticals to determine regional myocardial blood flow after intracoronary injection.

Myocardial Agents

Isotopes of K and Analogs

Myocardial cells, like all muscle cells, contain large amounts of potassium; reduced content of potassium in myocardial tissue has been shown to result from myocardial infarction or reduced coronary blood flow in dog and man (128, 129). The possibility of imaging the myocardial distribution of coronary blood flow using soluble tracer ions of elements of group I of the Periodic Table has been attempted by many workers with varying degree of success (130–137). The table gives the physical properties of several K, Rb, and Cs radionuclides proposed for myocardial scanning and myocardial clearance studies. The gamma energies of several of these are not suitable for imaging, and the radiation dose is relatively high for application in humans. In the last few years, successful production methods for obtaining ^{43}K and ^{129}Cs of high specific activity have been reported (138–143) and these have been used for static imaging of the myocardium.

Potassium-43 is produced in a nuclear reactor by bombarding ^{43}Ca with fast neutrons through the reaction ^{43}Ca(n, p)^{43}K using highly enriched ^{43}Ca (natural abundance 0.145%) (138). Alternatively, ^{43}K may be produced using a cyclotron through the following nuclear reactions, ^{40}Ar(α, p)^{43}K or ^{44}Ca(p, 2p)^{43}K, the former having been studied in detail (139, 140). However, the ^{42}K ($T_{1/2}$ = 12.4 hours, β^-_{max} = 3.52 MeV) contamination in these products is still a serious problem in terms of the increased patient radiation exposure. Myocardial perfusion imaging using ^{43}K has been performed in a large number of patient series (144).

Reactions for production of carrier-free ^{129}Cs are: (1) a combination of (p, 2n), and (p, 2p) reactions induced in enriched ^{130}Ba and (2) ^{127}I(α, 2n)^{129}Cs (142, 134). High production yields for this nuclide are reported.

Rubidium-81 (4.7 hour) decaying to short-lived ^{81m}Kr (13 seconds) may also find application for myocardial imaging by permitting multiple radiocardiography studies with low radiation exposure. ^{81}Rb could also be utilized by tagging to red blood cells (145, 146). The parent nuclide ^{81}Rb is produced through ^{79}Br (α, 2n) ^{81}Rb, or ^{81}Br (α, 4n) ^{81}Rb reactions induced by high-energy alpha particles or by ^{3}He particles at relatively low energies by ^{81}Br (^{3}He, 3n) ^{81}Rb, or ^{79}Br (^{3}He, n) ^{81}Rb reactions (99–101, 146). Elution of rubidium from the cells is reported to be negligible in 1 hour. The relatively short half-life of ^{81}Rb, combined with 511-keV annihilation radiation (26%) and 190-keV daughter radiation (65%) are quite useful for consideration of ^{81}Rb-tagged red cells for dynamic studies.

Other isotopes of K, specifically Rb and Cs, have also been used for myocardial-function and blood-flow studies using external precordial counting methods in many cases (147–156). The rate at which these isotopes are taken up by the myocardial tissue is determined by the rate of coronary blood flow and by the rate of exchange of the isotope between the circulating blood and the viable myocardial cell. Differences in the myocardial turnover characteristics of K, Rb, and Cs isotopes have been reported in animals (157, 158). Cognizance of these differences must be taken in the quantitation of myocardial blood flow studies and function measurements using these isotopes interchangeably.

Thallium-201 (Tl^+)

Thallium-201 has been used recently for myocardial visualization (159). This nuclide is prepared through the reaction ^{203}Tl $(p, 3n) \rightarrow$ ^{201}Pb (β^+, electron capture 9.4 hours) $\rightarrow$ ^{201}Tl. ^{201}Tl which decays by electron capture emits Hg x-rays (70 keV) and photons of 135 and 167 keV (2 and 8%, respectively) and has a convenient half-life of 74 hours (15).

^{13}N-Labeled NH^+_4

The use of ^{13}N ammonia for myocardial imaging has recently been reported (122, 160–162). The production of the short-lived ^{13}N ($T_{1/2} = 9.96$ minutes) as molecular nitrogen has been mentioned earlier. Nitrogen 13 labeled ammonia is produced in millicurie quantities through the nuclear reaction of deuterons on methane gas by the $^{12}C(d, n)^{13}N$ reaction (163) and passing the outgas first through a soda-lime column to remove traces of HCN and then bubbling into isotonic saline in which labeled ammonia dissolves. The ^{13}NH$_3$ dissolved in saline is of high purity and high radioactive concentration (> 10 mCi/cm^3). After intravenous injection of the pharmaceutical, the liver, myocardium, and urinary bladder are usually readily visualized. Blood disappearance of the ammonium ion is reported to be very rapid, with 85% of the activity leaving the blood in the first minute. The behavior of the ammonium ion seems to be markedly different from the potassium ion and its analogs (162).

Labeled Fatty Acids

It has been shown that long-chain fatty acids are specifically extracted by myocardial cells and serve as an important fuel in cellular oxidative metabolism (164, 165). Evans et al. (166) suggested the use of ^{131}I oleic acid bound to serum albumin for myocardial scanning. Fatty acids incorporating ^{123}I and ^{99m}Tc should be of great interest in terms of the application of these specific agents for myocardial imaging. Preliminary work has already been presented by Bonte and co-workers (167). Poe et al. (168) reported the myocardial extraction efficiency of various labeled fatty acids and carboxylates by direct coronary arterial injection. Chapter 10 was added as this text was going to press to indicate current promising work with ^{99m}Tc phosphates.

Nonspecific Agents

Diffusible and Nondiffusible Tracers

The radioactive isotopes of inert gases, ^{85}Kr and ^{133}Xe, are extensively used for the measurement of regional myocardial blood flow (12, 169-174). These gases are highly fat soluble and hence diffuse rapidly across capillary walls and cell membranes. The partition coefficients between myocardium and blood are 1.00 and 0.72 for Kr and Xe respectively (169). Solutions of xenon or krypton are injected directly into the coronary arteries through a catheter, and clearance of the radioactive material from the myocardium by precordial counting gives a measure of the myocardial perfusion. (Chapter 6, pp. 170-197).

^{99m}Tc-pertechnetate and ^{99m}Tc-serum albumin have been used for the measurement of regional blood flow in the myocardium by injecting a tracer bolus into the coronary artery (9, 175).

Agents Localizing in Infarcted Areas

Hydroxy mercury derivatives of fluorescein labeled with radioisotopes of mercury (176) and ^{99m}Tc-tetracycline (177) have been reported as agents accumulating in infarcted areas of the myocardium. These and similar agents which localize in the lesion and show up as a hot spot of increased uptake await detailed clinical investigations.

Radioactive Labeled Particles

The application of radioactive-labeled-particle techniques in the determination of regional blood flow (178) has been extended to the coronary circulation (179-184). Technetium-^{99m}-labeled macroaggregated human serum albumin (^{99m}Tc-MAA) particles of uniform size or ^{99m}Tc sulfur colloid macroaggregates or ^{99m}Tc-labeled human serum albumin microspheres are injected into the coronary circulation (10, 11, 184). Myocardial perfusion imaging is carried out in a manner similar to the technique of vascular microembolization with radioactive particles which is routinely done in human lung perfusion scanning.

Different methods of preparation of ^{99m}Tc macroaggregates or microspheres of human serum albumin are reported with high labeling efficiencies (185-188). Human albumin microspheres with a mean diameter of 22μ which can be tagged with ^{99m}Tc are available in a kit form commercially. ^{99m}Tc-sulfur colloid macroaggregates of uniform size (10 to 40μ diameter) are also prepared with nearly 100% labeling efficiency (189).

The effects of coronary arterial injection of macroaggregates or microspheres on coronary hemodynamics and myocardial function have been assessed by various workers (11, 183, 190, 191). It is concluded that slow coronary arterial injection of particles in the 10-60 micron range contained in 0.04 mg of albumin ($<200,000$ particles) and in small volume doses (0.1-0.5 ml) is safe and does not cause significant changes in coronary flow, myocardial contractility, arterial pressure and ventricular function. However, it has been shown that the intracoronary administration of ionic Cs$^+$ and ^{99m}Tc-labeled particles yield qualitatively similar myocardial scintigraphic images (192). These studies portend the use of these potassium analogs for clinical studies in preference to labeled particles which have an inherent hazard of arterial blockade.

REFERENCES

1. D. T. Mason, W. L. Ashburn, J. C. Harbert, L. S. Cohen, and E. Braunwald: Rapid Sequential Visualization of the Heart and Great Vessels in Man Using the Wide Field Anger Scintillation Camera. Radioisotope Angiography Following Injection of Technetium-99m, *Circulation* **39:** 19, 1969.
2. J. P. Kriss, L. P. Enright, W. G. Hayden, L. Wexler, and N. E. Shumway: Radioisotopic Angiocardiography. Wide Scope of Applicability in Diagnosis and Evaluation of Therapy in Diseases of the Heart and Great Vessels, *Circulation* **43:** 792, 1971.
3. L. Donato: Quantitative Radiocardiography and Myocardial Blood Flow Measurements with Radioisotopes, paper presented at Symposium on Dynamic Studies With Radioisotopes in Medicine, IAEA, Vienna, 1971.
4. Y. Ishii, and W. J. MacIntyre: Measurement of Heart Chamber Volumes by Analysis of Dilution Curves Simultaneously Recorded by Scintillation Camera, *Circulation* **44:** 37, 1971.
5. H. Wesselhoeft, P. J. Hurley, H. N. Wagner, Jr., and R. D. Rowe: Nuclear Angiocardiography in the Diagnosis of Congenital Heat Disease in Infants, *Circulation* **45:** 77, 1972.
6. D. Van Dyke, H. O. Anger, R. W. Sullivan, W. R. Vetter, Y. Yano, and H. G. Parker: Cardiac Evaluation from Radioisotope Dynamics, *J. Nucl. Med.* **13:** 585, 1972.
7. P. M. Weber, L. V. Dos Remedios, and I. A. Jasko: Quantitative Radioisotopic Angiocardiography, *J. Nucl. Med.* **13:** 815, 1972.
8. R. H. Jones, D. C. Sabiston, B. B. Bates, J. J. Morris, P. A. W. Anderson, and J. K. Goodrich: Quantitative Radionuclide Angiocardiography for Determination of Chamber to Chamber Cardiac Transit Times, *Am J. Cardiol.* **30:** 855, 1972.
9. E. E. Christensen, and F. J. Bonte: Radionuclide Coronary Angiography and Myocardial Blood Flow, *Radiology* **95:** 497, 1970.
10. W. L. Ashburn, E. Braunwald, A. L. Simon, K. L. Peterson, and J. H. Gault: Myocardial Perfusion Imaging with Radioactive-Labelled Particles Injected Directly into the Coronary Circulation of Patients with Coronary Artery Disease, *Circulation* **44:** 851, 1971
11. D. A. Weller, R. J. Adolph, H. N. Wellman, R. G. Carroll, and O. Kim: Myocardial Perfusion Scintigraphy after Intracoronary Injection of Tc-99m Labelled Albumin Microspheres. Toxicity and Efficacy for Detecting Myocardial Infarction in Dogs; Preliminary Results in Man, *Circulation* **46:** 963, 1972.
12. P. J. Cannon, R. B. Dell, and E. M. Dwyer, Jr.: Measurement of Regional Myocardial Perfusion in Man with Xe-133 and a Scintillation Camera, *J. Clin. Invest.* **51:** 964, 1972.
13. J. S. Laughlin, R. S. Tilbury, and J. R. Dahl: The Cyclotron; Source of Short Lived Radionuclides and Positron Emitters for Medicine, in *Recent Advances in Nuclear Medicine,* J. H. Lawrence, Ed., *Progress in Atomic Medicine,* Vol. 3. Grune and Stratton, New York, 1971.
13a. H. I. Glass: New Application of Radiopharmaceuticals Labelled with Cyclotron Produced Radionuclides, paper presented at Symposium on Medical Radioisotope Scintigraphy, IAEA, Monte Carlo, October 23–28, 1972.
14. H. N. Wagner, Jr., and H. Emmons: Characteristics of an Ideal Radiopharmaceutical, in *Radioactive Pharmaceuticals,* G. A. Andrews, R. M. Kniseley, and H. N. Wagner, Jr., Eds. U.S. Atomic Energy Commission, Washington, D.C., 1966.
15. C. M. Lederer, J. M. Hollander, and I. Perlman: *Table of Isotopes* 6th ed., Wiley, New York, 1967.
16. P. Richards: The Technetium—99m Generator, in *Radioactive Pharmaceuticals,* G. A. Andrews, R. M. Kniseley, and H. N. Wagner, Jr., Eds. U.S. Atomic Energy Commission, Washington, D.C., 1966.
17. H. S. Stern, I. Zolle, and J. G. McAfee: Preparation of Tc-99m Labelled Serum Albumin (Human), *Intern. J. Appl. Radiation Isotopes* **16:** 283, 1965.

18. R. B. R. Persson, and K. Lidén: Tc-99m Labelled Human Serum Albumin: a Study of the Labelling Procedure, *Intern J. Appl. Radiation Isotopes,* **20:** 241, 1969.
19. R. Dreyer, and R. Münze: Zur 99mTc-Markierung von Serumalbumin, Isotopen Praxis **5:** 296, 1969.
20. M. J. Williams, and T. Deegan: The Process Involved in the Binding of Technetium-99m to Human Serum Albumin, *Intern J. Appl. Radiation Isotopes* **22:** 767, 1971.
21. M. Lin, H. S. Winchell, and B. A. Shipley: Use of Fe(II) or Sn(II) Alone for Technetium Labelling of Albumin, *J. Nucl. Med.* **12:** 204, 1971.
22. W. C. Eckelman, G. Meinken, P. Richards: Tc-99m-Human Serum Albumin, *J. Nucl. Med.* **12:** 707, 1971.
23. P. P. Benjamin, A. Rejali, and H. Friedell: Electrolytic Complexation of Tc-99m at Constant Current: Its Applications in Nuclear Medicine, *J. Nucl. Med.* **11:** 147, 1970.
24. H. J. Dworkin, and R. F. Gutkowski: Rapid Closed System Production of Tc-99m-Albumin Using Electrolysis, *J. Nucl. Med.* **12:** 562, 1971.
25. W. C. Eckelman, G. Meinken, and P. Richards: High Specific Activity Tc-99m Human Serum Albumin, *Radiology* **102:** 185, 1972.
26. M. J. Williams, and T. Deegan: Tc-99m Labelled Serum Albumin in Cardiac Output and Blood Volume Studies, *Thorax* **26:** 460, 1971.
27. P. Richards: Personal Communication.
28. P. Hosain, F. Hosain, Q. M. Iqbal, N. Carulli, and H. N. Wagner, Jr.: Measurement of Plasma Volume Using Tc-99m, In-113m Labelled Proteins, *Brit. J. Radiol.* **42:** 627, 1969.
29. J. Fischer, R. Wolf, and A. Leon: Tc-99m as a Label for Erythrocytes, *J. Nucl. Med.* 8: 229, 1967.
30. U. Haubold, H. W. Pabst, and G. Hör: Scintigraphy of the Placenta with Tc-99m Labelled Erythrocytes, in *Medical Radionuclide Scintigraphy,* Vol. 2, IAEA, Salzburg, 1969.
31. R. Berger, and B. Johannsen: Markierung von Erythrozyten mit Tc-99m. Wiss. Z. Karl-Marx-Univ., Leipzig *Math-Naturwiss. R.* **18:** 635, 1969.
32. J. P. Novel, and P. Brunelle: Les Marquage des Hematies par le Technetium-99m, *Presse Med.* **78:** 73, 1970.
33. L. J. Anghileri, J. I. Lee, and E. S. Miller: The Tc-99m Labelling of Erythrocytes, *J. Nucl. Med.* **11:** 530, 1970.
34. G. L. Buraggi, R. Ringhini, A. Rodari, and D. Scaiano: Short-Living Radioisotopes (Tc-99m, In-113m) as a Label for Erythrocytes. Labelling Problems and Utilization in Spleen Studies, *J. Biol. Nucl. Med.* **15:** 23, 1971.
35. W. C. Eckelman, P. Richards, W. Hauser, and H. Atkins: Technetium Labelled Red Blood Cells, *J. Nucl. Med.* **12:** 22, 1971.
36. K. D. Schwartz, and M. Kruger: Improvement in Labelling Erythrocytes with Tc-99m Pertechnetate, *J. Nucl. Med.* **12:** 323, 1971.
37. J. McRae, and P. E. Valk: Alteration of Tc-99m Red Blood Cells, *J. Nucl. Med.* **13:** 399, 1972.
38. V. Korubin, M. N. Maisey, and P. A. McIntyre: Evaluation of Technetium Labelled Red Cells for Determination of Red Cell Volume in Man, *J. Nucl. Med.* **13:** 760, 1972.
39. M. B. Weinstein, O. I. Joensuu, P. Duffy, and B. Bennett: Technical Difficulties in Labelling Erythrocytes with Tc-99m: Identification of Agglutinating Substance, *J. Nucl. Med.* **12:** 183, 1971.
40. H. S. Stern, D. A. Goodwin, H. N. Wagner, Jr., and H. H. Kramer: In-113m a Short-Lived Isotope for Lung Scanning. *Nucleonics* **24**(10): 57, 1966.
41. D. A. Goodwin, H. S. Stern, H. N. Wagner, Jr., and H. H. Kramer: A New Radiopharmaceutical for Liver Scanning, *Nucleonics* **24**(11): 65, 1966.
42. H. S. Stern, D. A. Goodwin, U. Scheffel, H. N. Wagner, Jr., and H. H. Kramer: In-113m for Blood Pool and Brain Scanning, *Nucleonics* **25**(2): 62, 1967.

References

43. J. A. Burdine: Indium-113m Radiopharmaceuticals for Multipurpose Imaging, *Radiology* **93**: 605, 1969.
44. M. H. Adatepe, M. Welch, R. G. Evans, and E. J. Potchen: Clinical Applications of the Broad Spectrum Scanning Agent, Indium-113m, *Am. J. Roentgen Radium Therapy* **112**: 701, 1971.
45. F. Hosain, P. A. McIntyre, K. Poulouse, H. S. Stern, and H. N. Wagner, Jr.: Binding of Trace Amounts of Ionic Indium-113m to Plasma Transferrin, *Clin. Chim. Acta* **24**: 69, 1969.
46. R. D. Wochner, M. Adatepe, A. V. Amburg, and E. J. Potchen: A New Method for Estimation of Plasma Volume with the Use of the Distribution Space of Indium-113m Transferrin, *J. Lab. Clin. Med.* **75**: 711, 1970.
47. P. Hosain, P. Som, Q. M. Iqbal, and F. Hosain: Measurement of Cardiac Output with Indium-113m Labelled Transferrin, *Brit. J. Radiol.* **42**: 931, 1969.
48. H. Schicha, K. Vyska, V. Becker, L. E. Feinendegen, and L. Seipel: Minimal Cardiac Transit Times as Parameters of Cardiac Function. Measurements with In-113m and the Gamma Camera, in *Dynamic Studies with Radioisotopes in Medicine,* IAEA, Vienna, 1971.
49. R. E. O'Mara, G. Subramanian, J. G. McAfee, and C. L. Burger: Comparison of In-113m and Other Shortlived Agents for Cerebral Scanning, *J. Nucl. Med.* **10**: 18, 1969.
50. W. G. Myers, Radioisotopes of Iodine, in *Radioactive Pharmaceuticals,* G. A. Andrews, R. M. Kniseley, H. N. Wagner, Jr. eds., U.S. Atomic Energy Commission, Washington, D.C., 1966.
51. V. J. Sodd, and J. Blue: Cyclotron Generator of High Purity I-123, *J. Nucl. Med.* **9**: 349, 1968.
52. H. D. Hupf, J. S. Eldridge, and J. E. Beaver: Production of I-123 for Medical Applications, *Intern. J. Appl. Radiation Isotopes* **19**: 345, 1698.
53. D. J. Silvester, J. Sugden, and I. A. Watson: Preparation of I-123 by Particle Bombardment of Natural Antimony, *Radiochem. Radioanal. Letters* **2**: 17, 1969.
54. J. W. Blue, and V. J. Sodd: I-123 Production from the $\beta+$ Decay of Xe-123, in *Uses of Cyclotron in Chemistry, Metallurgy and Biology,* C. B. Amphlett, Ed., Butterworth, London, 1970.
55. V. J. Sodd, J. W. Blue, and K. L. Scholz: Pure ^{123}I Production with a Gas Flow Target, *J. Nucl. Med.* **12**: 395, 1971.
56. E. Lebowitz, M. W. Greene, and P. Richards: On the Production of I-123 for Medical Use, *Intern. J. Appl. Radiation Isotopes* **22**: 489, 1971.
57. M. A. Fusco, N. F. Peek, J. A. Jungerman, F. W. Zielinski, S. J. DeNardo, and G. L. DeNardo: Production of Carrier Free ^{123}I Using the ^{123}I $(p, 5n)$ ^{123}Xe Reaction, *J. Nucl. Med.* **13**: 729, 1972.
58. R. M. Lambrecht, and A. P. Wolf: The ^{122}Te $(^4$He, 3N) ^{123}Xe $\beta+$, E.C./2.1 hour ^{123}I Generator, *Radiation Res.* **52**: 32, 1972.
59. W. L. Hughes: The Chemistry of Iodination, *Ann. N.Y. Acad. Sci.* **70**: 3, 1957.
60. M. Tubis: Special Iodinated Compounds for Biology and Medicine, in *Radioactive Pharmaceuticals,* G. A. Andrews, R. M. Kniseley, and H. N. Wagner, Jr., Eds. U.S. Atomic Energy Commission, Washington, D.C., 1966.
61. A. S. McFarlane: Efficient Trace Labelling of Proteins with Iodine, *Nature* **182**: 53, 1958.
62. A. S. McFarlane: The Preparation of I-131 and I-125 Labelled Plasma Proteins, in *Radioisotope Techniques in the Study of Protein Metabolism,* Technical Report Series, No. 45, IAEA, Vienna, 1965.
63. W. F. Bale, R. W. Helmkamp, T. P. Davis, M. J. Izzo, R. L. Goodland, M. A. Contreras, and I. L. Spar: High Specific Activity Labelling of Proteins with I-131 by the IC1 Method, *Proc. Soc. Exptl. Biol. Med.* **122**: 407, 1966.
64. R. W. Helmkamp, M. A. Contreras, and W. F. Bale: I-131 Labelling of Proteins by the Iodine Monochloride Method, *Intern. J. Appl. Radiation Isotopes* **18**: 737, 1967.

65. R. W. Helmkamp, M. A. Contreras, and W. F. Bale: I-131 Labelling of Proteins at High Activity Level with ^{131}ICl Produced by Oxidation of Total Iodine in Na^{131}I Preparation, *Intern. J. Appl. Radiation Isotopes,* **18:** 747, 1967.
66. A. E. Reif: A Simple Procedure for High Efficiency Radioiodination of Proteins, *J. Nucl. Med.* **9:** 148, 1968.
67. E. Hallaba, H. El Asrag, and Y. Abou-Zeid: I-131 Labelling of Tyrosine by Iodine Monochloride, *Intern. J. Appl. Radiation Isotopes* **21:** 107, 1970.
68. E. Hallaba, and J. Drouet: I-131 Labelling of Human Serum Albumin by ICl and Chloramine-T Methods, *Intern J. Appl. Radiation Isotopes* **22:** 46, 1971.
69. U. Rosa: Electrochemical Labelling of Proteins with I-131, in *Proceedings of the Conference on Methods of Preparing and Storing Marked Molecules,* Euratom, Brussels, 1964.
70. W. M. Hunter, and F. C. Greenwood: Preparation of I-131 Labelled Human Growth Hormone of High Specific Activity, *Nature,* **194:** 495, 1962.
71. W. M. Hunter: Iodination of Protein Compounds, in *Radioactive Pharmaceuticals,* G. A. Andrews, R. M. Kniseley, and H. N. Wagner, Jr., Eds. U.S. Atomic Energy Commission, Washington, D.C.
72. J. J. Marchalonis: An Enzymatic Method for the Trace Iodination of Immunoglobulins and Other Proteins, *Biochem. J.* **113:** 299, 1969.
73. R. M. Lambrecht, C. Mantescu, C. Redvanly, and A. P. Wolf: Preparation of High Purity Carrier Free ^{123}ICl as Iodination Reagent for Synthesis of Radiopharmaceuticals, IV, *J. Nucl. Med.* **13:** 266, 1972.
74. H. N. Wellman, J. F. Mack, and E. L. Saenger: Study of the Parameters Influencing the Clincial Use of I-123, *J. Nucl. Med.* **10:** 381, 1969.
75. H. N. Wellman, and R. T. Anger, Jr.: Radioiodine Dosimetry and the Use of Radioiodine Other Than I-131 in Thyroid Diagnosis, *Seminars Nucl. Med.* **1:** 356, 1971.
76. P. Richards, E. Liebowitz, and L. G. Stang: The Brookhaven Linac Isotope Producer (BLIP), paper presented at Symposium on New Developments in Radiopharmaceuticals and Labelled Compounds, IAEA, Copenhagen, March 26–30, 1973.
77. J. F. Foster: Plasma Albumin, in *The Plasma Proteins,* Vol. 1, F. P. Putnam, Ed. Academic Press, New York, 1960.
78. P. Vernejoul, J. Del Olmo, R. H. Abundes, and C. Kellershohn: The Importance of Nuclear Indicators of Short Half Life in the Study of Cardiac Hemodynamics, *Nucl. Med.* **5:** 3, 1966.
79. P. Vernejoul, J. Valeyre, and C. Kellershohn: Dosimetry and Technique for the Use of ^{137m}Ba in Cardiac Hemodynamics, *Compt. Rend.* **264:** 10, 1967.
80. D. Ivancevic, P. Vernejoul, and C. Kellershohn: Right and Left Ventricular Volumes in Atrial Septal Defect Studied by Radiocardiography, paper presented at Symposium on Dynamic Studies with Radioisotopes in Medicine, IAEA, Vienna, 1971.
81. B. Delaloye, A. Bischof-Delaloye, W. Nedinger, and J. L. River: Les Explorations Dynamiques Cardiopulmonaires. Applications Cliniques du Radiocardiogramme Quantitatif et de la Circulographic Pulmonaire Realisee à la Camera à Scintillations, paper presented at *Symposium on Dynamic Studies with Radioisotopes in Medicine,* IAEA, Vienna, 1971.
82. M. Blau, R. Zielinski, and M. Bender: A Ba-137m Cow—a New Short-Lived Isotope Generator, *Nucleonics* **24**(10): 90, 1966.
83. J. J. Pinajian: A ^{137}Cs- ^{137m}Ba Isotope Generator, *J. Chem. Educ.* **44:** 212, 1967.
84. T. Nagai, and K. Watari: An Improved ^{137m}Ba Generator, *J. Nucl. Med.* **9:** 608, 1968.
85. F. P. Castronovo, Jr., R. C. Reba, and H. N. Wagner, Jr.: System for Sustained Intravenous Infusion of a Sterile Solution of ^{137m}Ba-EDTA, *J. Nucl. Med.* **10:** 242, 1969.
86. L. Donato: Studies of Cardiac and Pulmonary Function, in *Radioisotopes in Medical Diagnosis,* E. H. Belcher, and H. Vetter, Eds, Buttterworths, London, 1971.

87. P. D. Buckingham, and G. R. Forse: The Preparation and Processing of Radioactive Gases for Clinical Use, *Intern. J. Appl. Radiation Isotopes* **14:** 439, 1963.
88. C. M. E. Matthews, C. T. Dollery, J. C. Clark, and J. B. West: Radioactive Gases, in *Radioactive Pharmaceuticals,* G. A. Andrews, R. M. Kniseley, H. N. Wagner, Jr., Eds. U.S. Atomic Energy Commission, Washington, D.C., 1966.
89. M. D. Welch, and M. M. Ter-Pogossian: Preparation of Short-Lived Radioactive Gases for Medical Studies, *Radiation Res.* **36:** 580, 1968.
90. W. G. Myers, and W. W. Hunter, Jr.: C-11 in Bone and Lung Scanning, in *Symp. on Medical Radioisotope Scintigraphy,* Vol. II, IAEA, Vienna, 1969.
91. R. D. Finn, D. R. Christman, H. J. Ache, and A. P. Wolf: The Preparation of cyanide-^{11}C for Use in Synthesis of Organic Radiopharmaceuticals II, *Intern. J. Appl. Radiation Isotopes* **22:** 639, 1971.
93. H. I. Glass, A. Brant, J. C. Clark, A. C. DeGaretta, and L. G. Day: Measurement of Blood Volume Using Red Cells Labelled with Radioactive Carbon Monoxide, *J. Nucl. Med.* **9:** 571, 1968.
94. H. I. Glass, R. H. T. Edwards, A. C. DeGaretta, and J. C. Clark: ^{11}CO Red Cell Labelling for Blood Volume and Total Hemoglobin in Athletes: Effect of Training, *J. Appl. Physiol.* **26:** 131, 1969.
95. A. Hardweig, D. F. Rochester, and W. A. Briscoe: Measurement of Solubility Coefficients of Krypton in Water, Plasma and Human Blood Using Radioactive Kr-85, *J. Appl. Physiol.* **15:** 723, 1960.
96. S.-Y. Yeh, and R. E. Peterson: Solubility of Krypton and Xenon in Blood, Protein Solutions and Tissue Homogenates, *J. Appl. Physiol.* **20:** 1041, 1965.
97. K. Kitani: Solubility Coefficients of Kr-85, and Xe-133 in Water, Saline, Lipids and Blood, *Scand. J. Clin. Lab. Invest.* **29:** 127, 1972.
98. S.-Y. Yeh, and R. E. Peterson: Solubility of Carbon Dioxide, Krypton, and Xenon in Lipids, *J. Pharm. Sci.* **52:** 453, 1963.
99. T. Jones, and J. C. Clark: A Cyclotron Produced ^{81}Rb-^{81m}Kr Generator and Its Uses in Gamma Camera Studies, *Brit. J. Radiol.* **42:** 237, 1969.
100. T. Jones, J. C. Clark, J. M. Hughes, and D. Y. Rosenzweig: Kr-81m Generator and Its Uses in Cardiopulmonary Studies with the Scintillation Camera, *J. Nucl. Med.* **11:** 118, 1970.
101. Y. Yano, J. McRae, and H. O. Anger: Lung Function Studies Using Short Lived Kr-81m and the Scintillation Camera, *J. Nucl. Med.* **11:** 674, 1970.
102. N. Veall, and B. L. Mallett: The Partition of Trace Amounts of Xe Between Human Blood and Brain Tissues at 37°C, *Phys. Med. Biol.* **10:** 375, 1965.
103. J. Ladefoged, and A. M. Andersen: Solubility of Xe-133 at 37°C in Water, Saline, Olive Oil, Liquid Paraffin, Solutions of Albumin and Blood, *Phys. Med. Biol.* **12:** 353, 1967.
104. R. N. Arnot, J. C. Clark, and H. I. Glass: Investigation of Xe-127 as a Tracer for Measurement of Regional Cerebral Flow, in *Proceedings of 6th International Symposium on the Regulation of Cerebral Blood Flow (Cords)* R. W. Ross Russell, Ed., London, Pitman, 1970.
105. P. B. Hoffer, P. V. Harper, R. N. Beck, V. Stark, H. Krizek, L. Heck, and N. Lembares: Improved Xenon Images with Xe-127, *J. Nucl. Med* **14:** 172, 1973.
106. A. Liuzzi, J. Kearney, and G. Freedman: Use of Activated Charcoal for the Collection and Containment of Xe-133 Exhaled During Pulmonary Studies, *J. Nucl. Med.* **13:** 673, 1972.
107. K. E. Corrigan, J. Mantel, and H. H. Corrigan: Trapping System for Radioactive Gas, *Radiology* **96:** 571, 1970.
108. K. E. Corrigan, H. H. Corrigan, and J. Mantel: Recovery and Reconcentration of Radioactive Gases from the Cryogenic Trap, *Radiology* **106:** 615, 1973.

109. N. Veall: The Handling and Dispensing of Xe-133: Gas Shipments for Clinical Use, *Intern. J. Appl. Radiation Isotopes*, **16:** 385, 1965.
110. J. B. West: The Use of Radioactive Materials in the Study of Lung Function, *Medical Monograph Series, No. 1*, Radiochemical Centre, Amersham, U.K., 1966.
111. R. S. Tilbury, H. H. Kramer, and W. H. Wahl: Preparing and Dispensing Xe-133 in Isotonic Saline, *J. Nucl. Med.* **8:** 401, 1967.
112. K. Steimann, and N. Aspin: A Device for Dispensing Radioactive Xenon Gas, *Radiology* **92:** 396, 1969.
113. M. Loken, and G. S. Kush: Handling, Uses and Radiation Dosimetry of Xe-133, in *Medical Radionuclides, Radiation Dose and Effects*, U.S. Atomic Energy Commission, Washington, D.C., 1971.
114. P. S. Rummerfield, G. R. Jones, and W. L. Ashburn: Health Physics Aspects of Xe-133 Lung Studies, *Health Phys*, **12:** 547, 1971.
115. R. E. Snyder, and T. R. Overton: System for Handling and Dispensing Xe-133, *J. Nucl. Med.* **14:** 56, 1973.
116. J. Keany, A. Liuzzi, and G. S. Freedman: Large Dose Errors due to Redistribution of Xe-133 in Capsules and Plastic Syringes, *J. Nucl. Med.* **12:** 248, 1970.
117. A. LeBlanc, and P. C. Johnson: The Handling of Xe in Clinical Studies, *Phys. Med. Biol.* **16:** 105, 1971.
118. H. M. Abdel-Dayem: Handling of Radioactive Xe-133 Dissolved in Saline, *J. Nucl. Med.* **13:** 231, 1972.
119. J. C. Clark, C. M. E. Matthews, D. J. Silvester, and D. D. Vonberg: Using Cyclotron Produced Isotopes at Hammersmith Hospital, *Nucleonics* **25**(6): 54, 1967.
120. R. Green, B. Hoop, and H. Kazemi: Use of N-13 in Studies of Airway Closure and Regional Ventilation, *J. Nucl. Med.* **12:** 719, 1971.
121. P. D. Buckingham, and J. C. Clark: N-13 Solutions for Research Studies in Pulmonary Physiology, *Intern. J. Appl. Radiation Isotopes* **23:** 5, 1972.
122. B. Hoop, Jr., T. W. Smith, C. A. Burnham, J. E. Correll, G. L. Brownell, and C. A. Sanders: Myocardial Imaging with $^{13}NH_4+$ and a Multicrystal Positron Camera, *J. Nucl. Med.* **14:** 181, 1973.
123. C. T. Dollery, and J. B. West: Regional Uptake of Radioactive Oxygen, Carbon Monoxide and Carbon Dioxide in Lungs of Patients with Mitral Stenosis, *Circulation Res.* **8:** 765, 1960.
124. C. T. Dollery, J. B. West, D. E. L. Wilcken, J. F. Goodwin, and P. Hugh-Jones: Regional Pulmonary Blood Flow in Patients with Circulatory Shunts, *Brit. Heart J.* **23:** 225, 1961.
125. C. T. Dollery, P. Heimburg, and P. Hugh-Jones: Relationship between Blood Flow and Clearance Rate of Radioactive Carbon Dioxide and Oxygen in Normal and Edematous Lungs, *J. Physiol. (London)* **162:** 93, 1962.
126. H. Kazemi, F. Al Bazzaz, E. F. Parsons, and B. Hoop, Jr.: Distribution of Pulmonary Perfusion after Myocardial Infarction and Its Relationship to Arterial Hypoxemia, in *Dynamic Studies with Radioisotopes in Medicine*, IAEA, Vienna, 1971.
127. T. Jones, D. L. Levene, and R. Greene: Use of O-15 Labelled CO_2 for Inhalation Radiocardiograms and Measurement of Myocardial Perfusion, in *Dynamic Studies with Radioisotopes in Medicine*, IAEA, Vienna, 1971.
128. R. B. Jennings, H. M. Sommers, J. P. Kaltenbach, and J. J. West: Electrolyte Alterations in Acute Myocardial Ischemic Injury, *Circulation Res.* **14:** 260, 1964.
129. R. B. Case, M. G. Nasser, and R. S. Crampton: Biochemical Aspects of Early Myocardial Ischemia, *Am. J. Cardiol.* **24:** 766, 1969.
130. E. A. Carr, Jr., W. H. Beierwaltes, M. E. Patno, J. D. Bartlett, and A. V. Wegst: The Detection of Experimental Myocardial Infarcts by Photo Scanning, *Am. Heart J.* **64:** 650, 1962.

References

131. E. A. Carr, Jr., Beierwaltes, A. V. Wegst, and J. D. Bartlett: Myocardial Scanning with Rubidium-86, *J. Nucl. Med.* **3:** 76, 1962.
132. E. A. Carr, Jr., B. J. Walker, and J. D. Bartlett, Jr.: Diagnosis of Myocardial Infarcts after Administration of Cs-131, *J. Clin. Invest.* **42:** 922, 1963.
133. E. A. Carr, Jr., G. Gleason, J. Shaw, and B. Kroutz: Direct Diagnosis of Myocardial Infarction by Photoscan after Administration of Cs-131, *Am. Heart J.* **68:** 627, 1964.
134. J. T. McGeehan, A. Rodriguez-Antunez, and R. C. Lewis: Cs-131 Photoscan. Aid in the Diagnosis of Myocardial Infarction, *J. Am. Med. Assoc.* **204:** 585, 1968.
135. Y. Yano, and H. O. Anger: Visualization of Heart and Kidneys in Animals with Ultrashortlived Rb-82 and the Positron Scintillation Camera, *J. Nucl. Med.* **9:** 412, 1968.
136. G. E. Himes, L. V. Worth, and R. M. Smith: Cesium-131 Uptake and Distribution in the Human Heart; an Analysis of Cardiac Scans in 104 Patients, *J. Am. Osteopath. Assoc.* **64:** 575, 1965.
137. K. R. Bennett, R. O. Smith, P. H. Lehan, and H. K. Hellems: Correlation of K-42 Uptake with Coronary Arteriography, *Radiology* **102:** 117, 1972.
138. P. J. Hurley, M. Cooper, R. C. Reba, K. J. Poggenburg, and H. N. Wagner, Jr.: ^{43}KCl: a New Radiopharmaceutical for Imaging the Heart, *J. Nucl. Med.* **12:** 516, 1971.
139. J. C. Clark, M. L. Thakur, and I. A. Watson: The Production of K-43 for Medical Use, *Intern. J. Appl. Radiation Isotopes* **23:** 329, 1972.
140. H. H. Neely, N. S. MacDonald, J. Takahashi, J., R. Birdsall, and N. D. Poe: Production of K-43 with a Compact Cyclotron Using a Liquid Argon Target, *J. Nucl. Med. (Abstr.)* **14:** 433, 1973.
141. Y. Yano, D. Van Dyke, T. F. Budinger, H. O. Anger, and P. Chu: Myocardial Uptake Studies with ^{129}Cs and the Scintillation Camera, *J. Nucl. Med.* **11:** 663, 1970.
142. H. A. O'Brien, and H. B. Hupf: The Preparation of Carrier free ^{129}Cs, ^{132}Cs, and ^{136}Cs, *Intern. J. Appl. Radiation Isotopes* **22:** 95, 1971.
143. V. J. Sodd, J. W. Blue, and K. L. Scholz: ^{129}Cs Production via the ^{127}I $(\alpha, 2n)$ ^{129}Cs Reaction and Its Preparation as a Radiopharmaceutical, *Phys. Med. Biol.* **16:** 587, 1971.
144. B. L. Zaret, H. W. Strauss, N. D. Martin, H. P. Wells, and M. D. Flamm: Noninvasive Regional Myocardial Perfusion with Radioactive Potassium. Study of Patients at Rest, with Exercise and During Angina Pectoris, *N. Engl. J. Med.* **288:** 809, 1973.
145. J. B. Kahn, Jr.: The Entry of Rb into Human Erythrocytes, *J. Pharmacol. Exptl. Therap.* **136:** 197, 1962.
146. B. Friedmann, S. M. Lewis, H. I. Glass, L. Szur, I. A. Watson: Labelling of Red Blood Cells with ^{81}Rb for Spleen Scans, *Brit. J. Radiol.* **41:** 815, 1968.
147. W. D. Love, and G. E. Burch: Estimation of the Ratio of Uptake of Rb-86 by the Heart, Liver, and Skeletal Muscle of Man With and Without Cardiac Disease, *J. Appl. Radiation Isotopes* **3:** 207, 1955.
148. W. D. Love, and G. E. Burch: Influence of the Rate of Coronary Plasma Flow on the Extraction of ^{86}Rb from Coronary Blood, *Circulation Res.* **7:** 24, 1959.
149. M. M. Levy, and J. M. DeOliveira: Regional Distribution of Myocardial Blood Flow in the Dog as Determined by Rb-86, *Circulation Res.* **9:** 96, 1961.
150. L. Donato, G. Bartolomei, and R. Giodani: Evaluation of Myocardial Blood Perfusion in Man with Radioactive Potassium or Rubidium and Precordial Counting, *Circulation* **29:** 195, 1964.
151. L. Donato, G. Bartolomei, G. Federighi, and G. Torreggiani: Measurement of Coronary Blood Flow by External Counting with Radioactive Rubidium. Critical Appraisal and Validation of the Method, *Circulation* **33:** 708, 1966.
152. A. Cohen, E. J. Zaleski, H. Baleiron, T. B. Stock, C. Chiba, and R. J. Bing: Measurement of Coronary Blood Flow Using ^{84}Rb and the Coincidence Counting Method. A Critical Analysis. *Am. J. Cardiol.* **19:** 556, 1967.

153. P. L. McHenry, and S. B. Knoebel: Measurement of Coronary Blood Flow by Coincidence Counting and a Bolus of ^{84}RbCl, *J. Appl. Physiol.* **22**: 495, 1967.
154. W. D. Love, R. O. Smith, and P. E. Pulley: Mapping Myocardial Mass and Regional Coronary Blood Flow by External Monitoring of ^{42}K and ^{86}Rb Clearance, *J. Nucl. Med.* **10**: 702, 1969.
155. D. Boettcher, G. Corsini, C. D. Daniels, C. Cowan, and R. J. Bing, Determination of Myocardial Blood Flow in the Anesthetized Dog After a Bolus of ^{84}Rb, *J. Nucl. Med.* **10**: 83, 1969.
156. A. Maseri, A. Pesola, T. Duce, and L. Donato: Myocardial Blood Flow Measurements by Radioactive Rubidium and Potassium, Importance and Correct Estimate of Blood Contribution Without the Use of an Intravascular Tracer, *J. Nucl. Biol. Med.* **15**(3): 87, 1971.
157. W. D. Love, Y. Ishihara, L. D. Lyon, and R. O. Smith: Differences in the Relationships Between Coronary Blood Flow and Myocardial Clearance of Isotopes of K, Rb and Cs, *Am. Heart J.* **76**: 353, 1968.
158. N. D. Poe: Comparative Myocardial Uptake and Clearance Characteristics of K and Cs, *J. Nucl. Med.* **13**: 557, 1972.
159. E. Lebowitz, M. W. Greene, P. Bradley-Moore, H. Atkins, A. Ansari, P. Richards, and E. Belgrave: Thallium-201 for Medical Use, *J. Nucl. Med.* **14**: 421, 1973.
160. W. W. Hunter, and W. G. Monahan: ^{13}N Ammonia: A New Physiologic Radiotracer for Molecular Medicine, *J. Nucl. Med.* **12**: 368, 1971.
161. W. G. Monahan, R. S. Tilbury, and J. S. Laughlin: Uptake of ^{13}N-Labelled Ammonia, *J. Nucl. Med.* **13**: 274, 1972.
162. P. V. Harper, K. A. Lathrop, H. Krizek, N. Lembares, V. Stark, P. B. Hoffer: Clinical Feasibility of Myocardial Imaging with ^{13}NH$_3$, *J. Nucl. Med.* **13**: 278, 1972.
163. R. S. Tilbury, J. R. Dahl, W. G. Monahan, and J. S. Laughlin: The Production of ^{13}N Labelled Ammonia for Medical Use, *Radiochem. Radioanal. Letters* **8**: 317, 1971.
164. A. Carlsten, B. Hallgren, R. Jagenburg, A. Svanborg, and L. Werko: Myocardial Metabolism of Glucose, Lactic Acid, Amino Acids and Fatty Acids in Healthy Human Individuals at Rest and at Different Work Loads, *Scand. J. Clin. Lab. Invest.* **13**: 418, 1961.
165. J. R. Evans, L. H. Opie, and J. C. Shipp: Metabolism of Palmitic Acid in Perfused Rat Heart, *Am. J. Physiol.* **205**: 766, 1963.
166. J. R. Evans, R. W. Gunton, R. G. Baker, D. S. Beanlands, and J. C. Spears: Use of Radioiodinated Fatty Acid for Photoscans of the Heart, *Circulation Res.* **16**: 1, 1965.
167. F. J. Bonte, K. D. Graham, J. G. Moore, R. W. Parkey, and G. C. Curry: Preparation of ^{99m}Tc Oleic Acid Complex for Myocardial Imaging, *J. Nucl. Med.* (Abstr.) **14**: 381, 1973.
168. N. D. Poe, G. D. Robinson, and N. S. MacDonald: Myocardial Extraction of Variously Labelled Fatty Acids and Carboxylates. *J. Nucl. Med.* **14**: 440, 1973.
169. R. S. Ross, K. Ueda, P. R. Lichtlen, J. R. Rees: Measurement of Myocardial Blood Flow in Animals and Man by Selective Injection of Radioactive Inert Gas into the Coronary Arteries, *Circulation Res.* **15**: 28, 1964.
170. E. Linder: Measurements of Normal and Collateral Coronary Blood Flow by Close Arterial and Intramyocardial Injection of Kr-85 and Xe-133, *Acta Physiol. Scand.* **272**: 5, 1966.
171. J. M. Sullivan, W. J. Taylor, W. C. Elliot, and R. Gorlin: Regional Myocardial Blood Flow, *J. Clin. Invest.* **46**: 1402, 1967.
172. J. B. Bassingthwaighte, T. Strandell, and D. E. Donald: Estimation of Coronary Blood Flow by Washout of Diffusible Indicators, *Circulation Res.* **23**: 259, 1968.
173. F. J. Klocke, and S. M. Wittenberg: Heterogeneity of Coronary Blood Flow in Human Coronary Artery Disease and Experimental Myocardial Infarction, *Am. J. Cardiol.* **25**: 782, 1969.

174. A. Pitt, G. C. Friesinger, and R. S. Ross: Measurement of Blood Flow in the Right and Left Coronary Artery Beds in Humans and Dogs Using Xe-133 Technique, *Cardiovasc. Res.* **3**: 100, 1969.
175. W. Hayden, L. Wexler, and J. P. Kriss: Radionuclide Coronary Arteriography in Man, *J. Nucl. Med.* **13**: 435, 1972.
176. B. Vavrejn, P. Malek, J. Ratusky, L. Kronrad, and J. Kolc: Detection of Myocardial Infarcts by *in-vivo* Scanning Using Mercurascan, in *Medical Radioisotope Scintigraphy*, Vol. 2, IAEA, Salzberg, 1969.
177. B. L. Holman, M. K. Dewanjee, J. Idoine, C. P. Fliegel, M. A. Davis, S. Treves, and P. Eldh: Detection and Localization of Experimental Myocardial Infarction with Tc-99m Tetracycline, *J. Nucl. Med.* in Press.
178. H. N. Wagner, Jr., B. A. Rhodes, Y. Sasaki, and J. P. Ryan: Studies of the Circulation with Radioactive Microspheres, *Invest. Radiol.* **4**: 374, 1969.
179. H. Ueda, S. Kaihara, K. Ueda, Y. Sugushita, Y. Sasaki, and M. Iio: Regional Myocardial Blood Flow Measured by I-131 Labelled Macroaggregated Albumin (I-131 MAA), *Japan. Heart J.* **6**: 534, 1965.
180. J. L. Quinn, III, M. Serratto, and P. Kezdi: Coronary Artery Bed Photoscanning Using Radioiodine Albumin Macroaggregates (RAMA), *J. Nucl. Med.* **7**: 107, 1966.
181. R. J. Domenech, J. I. Hoffman, M. I. H. Nobel, K. Saunders, J. R. Henson, and S. Subijanto: Total and Regional Coronary Blood Flow Measured by Radioactive Microspheres in Conscious and Anesthetized Dogs, *Circulation Res.* **25**: 581, 1969.
182. M. Endo, T. Yamazaki, S. Konno, H. Hiratsuka, T. Akimoto, T. Tanaki, and S. Sakakibara: The Direct Diagnosis of Human Myocardial Ischemia Using ^{131}I-MAA via the Selective Coronary Catheter, *Am. Heart J.* **80**: 498, 1970.
183. N. Fortuin, S. Kaibara, L. Becker, and B. Pitt: Regional Myocardial Blood Flow in Dog Studies with Radioactive Microspheres, *Cardiovasc. Res.* **5**: 331, 1971.
184. H. R. Schelbert, W. L. Ashburn, J. W. Covell, A. L. Simon, E. Braunwald, and J. Ross, Jr.: Feasibility and Hazards of the Intracoronary Injection of Radioactive Serum Albumin Macroaggregates for External Myocardial Perfusion Imaging, *Invest. Radiol.* **6**: 379, 1971.
185. M. M. Gwyther, and E. O. Field: Aggregated ^{99m}Tc-Labelled Albumin for Lung Scintiscanning, *Intern. J. Appl. Radiation Isotopes* **17**: 485, 1966.
186. I. Zolle, B. A. Rhodes, and H. N. Wagner Jr.: Preparation of Metabolizable Radioactive Human Serum Albumin Microspheres for Studies of the Circulation, *Intern. J. Appl. Radiation Isotopes* **21**: 155, 1970.
187. J. A. Burdine, R. E. Sonnemaker, L. A. Ryder, and H. J. Spjut: Perfusion Studies with Technetium-99m Human Albumin Microspheres, *Radiology* **95**: 101, 1970.
188. B. A. Rhodes, H. S. Stern, J. A. Buchanan, I. Zolle, and H. N. Wagner, Jr.: Lung Scanning with ^{99m}Tc-Microspheres, *Radiology* **99**: 613, 1971.
189. V. Ficken, S. Halpern, C. Smith, Jr., L. Miller, and C. Bogardus, Jr.: ^{99m}Tc-Sulfur Colloid Macroaggregates, *Radiology* **97**: 289, 1970.
190. N. D. Poe: The Effects of Coronary Arterial Injection of Radioalbumin Macroaggregates on Coronary Hemodynamics and Myocardial Function, *J. Nucl. Med.* **12**: 724, 1971.
191. D. R. Richmond, T. Yipintsoi, C. M. Coulam, J. L. Titus, and J. B. Bassingthwaighte: Macroaggregated Albumin Studies of the Coronary Circulation in the Dog, *J. Nucl. Med.* **14**: 129, 1973.
192. N. D. Poe: Comparative Myocardial Distribution Patterns after Intracoronary Injection of Cs and Labelled Particles, *Radiology* **106**: 341, 1973.

CONTRIBUTING EDITORS: ROBERT W. PARKEY

FREDERICK J. BONTE

10. Imaging of Acute Myocardial Infarctions

The role of nuclear medicine in determining the presence, location and size of myocardial infarctions is rapidly changing; it is no longer simply a research endeavor, but has become a useful clinical tool. In early attempts to detect myocardial infarctions agents were used that labeled the normal myocardium, leaving the abnormal myocardium as a relative void when imaging. Recently, however, several radionuclide tracers have been developed that have selective affinity for ischemic or necrotic myocardium, providing a positive image at the site of the lesion, and, therefore, a more effective method of imaging.

The first attempts to label the normal myocardium involved the use of the predominantly intracellular cation potassium and its analogues (1.4). Of the cations, ^{43}K (5,6) and ^{129}Cs (7) have shown clinical usefulness. These cations label perfused tissue and an area of decreased or absent uptake indicates the presence of underperfused or nonperfused myocardium. Thus they do not distinguish between acutely and chronically ischemic regions. This, coupled with the marginal count rates and nonideal collimation characteristics of these high energy photons, presents a disadvantage in determining the size or stage of an acute myocardial infarction. However Strauss et al. (8) have used this technique to advantage to evaluate areas of transient myocardial ischemia before and after exercise using ^{43}K. The basic technique of myocardial scintigraphy with ^{43}K is described in detail on page 157. In most patients the maximum myocardial concentration of potassium is achieved after 10 to 30 minutes.

Other radiopharmaceuticals used to label normal myocardium are ammonium labeled with cyclotron-produced ^{13}N (9) and radio-iodinated fatty acids. Substances injected into the coronary arteries to measure myocardial perfusion, such as ^{133}Xe (10-13) and radiolabeled microspheres (14,15), can demonstrate a myocardial infarction as an area of decreased activity, and as a delay in washout in the case of radioactive gases, but they also do not distinguish between acutely and chronically ischemic regions.

All the radionuclide techniques mentioned above label the normally perfused myocardium leaving the underperfused myocardium (areas of ischemia, acute or chronic infarction) as relative voids. A more effective method of imaging would be to label only the abnormal tissue, leaving the normal tissue unlabeled. This has been accomplished in patients with acute myocardial infarction by using ^{99m}Tc-phosphate, ^{99m}Tc-tetracycline, and ^{67}Ga citrate.

Holman et al. (16,17) have shown that ^{99m}Tc-stannous tetracycline is deposited in acutely infarcted myocardium; results in humans showed good correlation between electrocardiographic and scintigraphic localization, and correlation between the size of the infarct by scintigraphy and the maximum serum creatine phosphokinase activity. The clinical usefulness of ^{99m}Tc (Sn) tetracycline in detecting acute myocardial infarctions is limited because (1) uptake in the liver is high, resulting in difficulty in separating diaphragmatic infarcts from liver, and (2) clearance of tetracycline from the blood is slow, leading to a considerably later time of maximal myocardial tracer concentration relative to plasma (24 to 48 hours). Because of the 6-hour half-life of ^{99m}Tc, the empirically derived optimal imaging time is 24 hours after injection.

Kramer et al. (18) have shown that gallium-67 accumulates in regions of acute myocardial infarction, probably in the associated inflammatory cell response to the ischemic tissue. However, ^{67}Ga uptake occurred in only five of eight hearts with documented infarction. The interval between infarction and scanning varied, and an optimal time for gallium scanning has not been established. This, along with the activity in the liver, may render ^{67}Ga relatively insensitive as a method of detecting acute myocardial infarction.

The best images to date (September 1974) of acute myocardial infarctions have been obtained using ^{99m}Tc-stannous pyrophosphate, a standard bone scanning agent (19-20). D'Agostino (21) described the localization of calcium ions within the mitochondria of myocardial cells in steroid-induced focal necrosis. D'Agostino and Chiga (22) observed that the calcium seemed to be incorporated within a crystalline structure which they believed to be hydroxyapatite. Shen and Jennings (23,24) approached this problem with ^{45}CaCl$_2$ tracer studies, giving the agent to dogs at selected intervals before and after occlusion of major branches of the coronary arteries. They drew two conclusions from their study: (1) calcium uptake is a feature of irreversible cellular injury that occurs only when arterial blood flow is present, and (2) calcium uptake behaves as if it were an active process associated with mitochondrial accumulation of calcium into what Shen and Jennings believe to be intramitochondrial granules of calcium phosphate. However, whether the structures formed within the mitochondria represent hydroxyapatite or some other molecule that contains calcium and phosphate, they are very likely the site of localization not only of calcium ions, but of the bone-imaging tracers such as ^{99m}Tc-stannous pyrophosphate. Many laboratories are carrying out further studies to verify the localization of the ^{99m}Tc-stannous pyrophosphate, and to consider other less likely mechanisms such as localization in inflammatory tissue.

The distribution of other bone-imaging tracers (25) was investigated, and it was shown that ^{99m}Tc-stannous polyphosphate, and ^{99m}Tc-stannous 1, hydroxy-ethylidene-1, 1-disodium phosphonate (diphosphonate) are potentially useful in myocardial infarct imaging. Fluorine 18 showed uptake in infarcted tissue, but simultaneous bone uptake occurs sufficiently early to decrease its usefulness.

^{99m}Tc-stannous pyrophosphate (PYP) imaging studies show that experimental infarctions become visible on scintillation camera images when the interval after infarction is 12 to 16 hours, and the interval from administration of dose to imaging is from 30 to 60 minutes. Localization continues to occur for 4 to 6 days, but begins to fade thereafter, and selective concentration in the infarct usually disappears by the fourteenth post infarction day. In humans, positive images have been seen 24 hours after infarction; intensity of uptake improves slightly over the next 48 hours, and begins to fade after 6 to 8 days. Unlike the animal model, some human infarcts have remained faintly positive for several weeks.

The technique for imaging acute myocardial infarction with ^{99m}Tc-PYP is as follows: a dose of 15 mCi ^{99m}Tc tagged to 5 mg of stannous pyrophosphate is injected intravenously. Scintigrams are obtained from 45 to 80 minutes after injection in the anterior, lateral, and one or more left anterior oblique projections.

The scintigrams are graded zero to 4+, depending on the activity over the myocardium. Zero represents no activity; 1+ indicates questionable activity; 2+ definite activity; 3+ and 4+ represent increasing degrees of activity within the infarct. Zero and 1+ are considered negative and 2+, 3+, and 4+ are considered positive. This system is arbitrary, but when used clinically it has shown good correlation with electrocardiographic and enzymatic criteria both for site and size of infarctions in 300 cases studied.

Visualization of a volume of damaged myocardium related to transmural infarction is graded as 3 to 4+ positive (20-26) and subendocardial infarction as 2+ (27), indicating varying degrees of intensity of tracer uptake as seen on imaging. (The reader is referred to this literature for illustrations of the different grades and locations of infarctions.)

Although computer processing of the images is not usually necessary to visualize an infarct, it allows greater precision in determining size and exact location. Simple background subtraction and contrast enhancement usually improve infarct definition. The rib structures that are prominently visualized with bone scanning agents can be removed from the image by using a one-dimensional, recursive, band-reject digital filter (28).

This technique represents a simple, noninvasive, and safe procedure for documenting the presence and location of acute myocardial infarctions when the scintigrams are performed within the first 6 days of symptoms suggesting infarction. Detection of extension of an infarction and estimation of infarct size and location are additional benefits. Optimal results are obtained if from 1 to 3 days pass after infarction and prior to imaging. Disease processes in which a false positive image may occur are breast tumors, functioning breast parenchyma in premenopausal females, healing rib fractures or other rib abnormalities, or trauma to chest wall muscles. ^{99m}Tc-pyrophosphate is rapidly cleared by the kidneys, so that little blood pool radioactivity is present in the heart 1 hour after injection. If a radiopharmaceutical is cleared from the blood slowly, or if renal disease prevents prompt clearance, retained radioactivity in the cardiac blood pool at the time of imaging may give rise to false positive scintigrams. The most common cause for false negative scintigrams is delay in performance of the imaging procedure until after the sixth or seventh day postinfarction, when the affinity of infarcted tissue for the tracer is rapidly decreasing.

Selective imaging of acute myocardial infarction by means of a noninvasive method may quickly become established as an area of competence in clinical nuclear medicine. Better radiopharmaceuticals are likely to be revealed with further study. At present, ^{99m}Tc tagged phosphates provide images of adequate quality in a clinical environment.

REFERENCES

1. E. A. Carr, Jr., W. H. Beierwaltes, A. V. Wegst, and J. D. Barlett, Jr.: Myocardial Scanning with Rubidum-86, *J. Nucl. Med.* **3:** 76, 1962.
2. E. A. Carr, Jr., G. Gleason, J. Shaw, and B. Krontz: Direct Diagnosis of Myocardial Infarction by Photoscanning after the Administration of Cesium-131, *Am. Heart J.,* **68:** 627–636, 1964.
3. D. Nolting, R. Mack, E. Luthy, M. Kirsch, and C. Hogancamp: Measurement of Coronary Blood Flow and Myocardial Rubidium Uptake with Rb-86, *J. Clin. Invest.* **37:** 921, 1958.
4. W. D. Love, R. O. Smith, and P. E. Pulley: Mapping Myocardial Mass and Regional Coronary Blood Flow by External Monitoring of K-42 or Rb-86 Clearance, *J. Nuclear Med.* **10:** 702–707, 1969.
5. P. J. Hurley, M. Cooper, R. C. Reba, K. J. Poggenburg, and H. N. Wagner, Jr.: ^{43}KCl: A New Radiopharmaceutical for Imaging the Heart, *J. Nuc. Med.* **12:** 516–519, 1971.
6. H. N. Wagner, Jr., Personal communication, 1972.
7. D. W. Romhilt, R. J. Adolph, V. J. Sodd, N. I. Levenson, L. S. August, H. Nishiyama, and R. A. Berke: Cesium-129 Myocardial Scintigraphy to Detect Myocardial Infarction, *Circulation* **48:** 1242–1251, 1973.
8. H. W. Strauss, B. L. Zaret, N. D. Martin, H. P. Wells, Jr., and M. D. Flamm, Jr.: Noninvasive Evaluation of Regional Myocardial Perfusion with Potassium-43. Technique in Patients with Exercise-Induced Transient Myocardial Ischemia, *Radiology* **108:** 85, 1973.
9. P. V. Harper, K. A. Lathrop, H. Krizek, N. Lembares, V. Stark, and P. B. Hoffer: Clinical Feasibility of Myocardial Imaging with ^{13}NH$_3$, *J. Nucl. Med.* **13:** 278–280, 1972.
10. P. J. Cannon, R. B. Dell, and E. M. Dwyer, Jr.: Measurement of Regional Myocardial Perfusion in Man with ^{133}Xenon and a Scintillation Camera, *J. Clin. Invest.* **51:** 964–977, 1972.
11. P. J. Cannon, R. B. Dell, and E. M. Dwyer, Jr.: Regional Myocardial Perfusion Rates in Patients with Coronary Artery Disease, *J. Clin. Invest.* **51:** 978–994, 1972.
12. R. W. Parkey, E. M. Stokely, S. E. Lewis, and F. J. Bonte: Compartmental Analysis of the ^{133}Xe Regional Myocardial Blood-Flow Curve, *Radiology* **104:** 425–426, 1972.
13. B. L. Holman, D. Jewitt, P. F. Cohn, et al.: The Effect of Rapid Atrial Pacing on Regional Myocardial Blood Flow, *Circulation* **48:** Suppl. 4:64, 1973.
14. W. Ashburn, E. Braunwald, A. Simon, et al.: Myocardial Perfusion Imaging with Radioactive-Labeled Particles Injected Directly into the Coronary Circulation of Patients with Coronary Artery Disease, *Circulation* **44:** 851–865, 1971.
15. D. A. Weller, R. J. Adolph, H. N. Wellman, et al.: Myocardial Perfusion Scintigraphy after Intracoronary Injection of ^{99m}Tc-Labeled Human Albumin Microspheres, *Circulation* **46:** 963–975, 1972.
16. F. L. Holman, M. K. Dewanjee, J. Idoine, C. P. Fliegel, M. A. Davis, S. Treves, and P. Eldh: Detection and Localization of Experimental Myocardial Infarction with ^{99m}Tc-Tetracycline, *J. Nucl. Med.* **14:** 595–599, 1973.
17. B. L. Holman, M. Lesch, F. G. Zweiman, J. Temte, B. Lown, and R. Gorlin: Detection and Sizing of Acute Myocardial Infarcts with ^{99m}Tc(Sn) Tetracycline, *N. Engl. J. Med.* **291:** 159–163, 1974.

18. R. J. Kramer, R. E. Goldstein, J. W. Hirshfeld, W. C. Roberts, G. S. Johnston, and S. E. Epstein: Accumulation of Gallium-67 in Regions of Acute Myocardial Infarction, *Am. J. Cardiol.* **33:** 861-867, 1974.
19. F. J. Bonte, R. W. Parkey, K. D. Graham, J. Moore, and E. M. Stokely: A New Method for Radionuclide Imaging of Myocardial Infarcts, *Radiology* **110:** 473-475, 1974.
20. R. W. Parkey, F. J. Bonte, S. L. Meyer, J. M. Atkins, J. L. Willerson, and G. C. Curry: A New Method for Radionuclide Imaging of Acute Myocardial Infarction in Humans, *Circulation* in press (Sept. 1974).
21. A. N. D'Agostino: An Electron Microscopic Study of Cardiac Necrosis Produced by α-Fluorocortisol and Sodium Phosphate. *Am. J. Pathol.* **45:** 633-644, 1964.
22. A. N. D'Agostino, and M. Chiga: Mitochondrial Mineralization in Human Myocardium, *Am. J. Clin. Pathol.* **53:** 829-824, 1970.
23. A. C. Shen, and R. B. Jennings: Myocardial Calcium and Magnesium in Acute Ischemic Injury, *Am. J. Pathol.* **67:** 417-440, 1972.
24. A. C. Shen, and R. B. Jennings: Kinetics of Calcium Accumulation in Acute Myocardial Ischemic Injury, *Am. J. Pathol.* **67:** 441-452, 1972.
25. F. J. Bonte, R. W. Parkey, K. D. Graham, and J. G. Moore: Distribution of Several Agents Useful in Imaging Myocardial Infarcts, submitted *J. Nucl. Med.*, July 1974.
26. J. T. Willerson, R. W. Parkey, F. J. Bonte, S. L. Meyer, J. M. Atkins, and E. M. Stokely: A New Method for Recognizing Acute Myocardial Infarction in Patients, submitted to *N. Engl. J. Med.* Aug. 1974.
27. J. T. Willerson, R. W. Parkey, F. J. Bonte, S. L. Meyer, and E. M. Stokely: Acute Subendocardial Myocardial Infarction in Patients: Its Detection by Technetium 99-m Stannous Pyrophosphate Myocardial Scintigrams, submitted to *Circulation* Aug. 1974.
28. E. M. Stokely, R. W. Parkey, S. E. Lewis, and F. J. Bonte: ^{99m}Tc Stannous Pyrophosphate Myocardial Scintigram Processing by Digital Computer, submitted *J. Nucl. Med.*, July 1974.

Index

Access, random, 236
  sequential, 236
Activated charcoal, 263
Adrenergic state, 107
Akinesis, 90, 93, 167, 179
Albumin, human serum, 131$_I$,
    136, 256
  microspheres, 113, 268
  ^{99m}Tc, 6, 26, 125, 131, 163, 256
Albumin macroaggregates, 163
  uniformity, 164
Analog buffering, 238
Analog-to-digital converter, 232, 235, 238
Aneurysm, 90, 184
  aortic, 83, 93
  left ventricular, 74, 83, 90, 93
  paradoxical myocardial kinetics, 90
  pulmonary artery, 97
Anger scintillation camera, 12, see also Scintillation camera
Angina, 156, 196
Angiography, 4, 5, 124, 179
  valvular lesions, 84
Annihilation radiation, 255
Anomalous pulmonary venous drainage, 37, 40
Antipyrine, 192
Aorta, 6, 15, 97
Aortic stenosis, 181
  ejection fraction increase, 74
  poststenotic dilatation, 74
Aortic valve, 8
Area of interest, see Region of interest
Area-volume, 132
Arteriovenous shunt, 172; see also Shunt

Atrial septal defect, 147; see also Congenital heart disease, shunt
Atrium, left, 5, 6, 8, 66, 82ff, 131, 209
Atrium, right, 5, 15, 69, 73, 84, 88, 92, 108, 115, 209
Autofluoroscope, 12, 19, 66, 174, 249
AV fistula, 107
Average life, 255

Background, 35, 128, 138, 218, 242, 256
  subtraction, 9, 131
^{137m}Barium, 262
Blood-pool measurement, 256
Bolus, 5, 23, 29, 41, 125, 137, 173, 194, 226, 233, 237, 256
  transit, 128; see also transit time
Buffer storage, 235, 243
Bypass graft, 164

Calcium, 279
Carbon monoxide, 263; see also CO
Cardiac catheterization, see Catheterization
Cardiac fat, 192
Cardiac motion, 194, 195, 280
Cardiac output, 1, 4, 6, 28, 137, 138, 213, 256, 260
Cardiac tumors, 99
  left atrial myxoma, 99
  metastatic, 101
Cardiomyopathy, 97, 144
Catheterization, 4, 5, 6, 32, 36, 41, 119, 150, 155, 163, 164, 167, 172, 175, 177, 180, 193
Cavity dimensions, 90, 230
Cesium, see Cs

283

Chambers, cardiac, 4, 230; see also Atrial ventricle
Chambers overlap, 242; see also Background
Chamber volume, 26
Children, 5
Circulation time, 5; see also Transit time
Cloud-chamber, 136
11_{CO}, 262
CO, see Carbon Monoxide, Cardiac output
Coarctation of the aorta, 35, 37
Coincidence-summing, 250
Collateral blood flow, 186, 194, 196
Collimator, 12, 157, 174, 195, 249
 converging, 5
 pinhole, 5
 septal absorption, 124
Compartmental analysis, 191, 206, 228
Compton scattering, 128
Computer, 2, 12
 analog, 9, 27
 buffer storage, 243
 core, 235
 digital, 136, 176, 233
 interface, 233
 processing of image, 280
 scintillation camera, 66
 software, see Programming
Congenital heart disease, 32, 56
 coarctation of the aorta, 35, 37, 52
 collateral circulation, bronchial, 37
 Ebstein's anomaly, 56
 pulmonary stenosis, 40, 56
 shunt, 6, 35
  anomalous pulmonary venous drainage, 40
  aorta-pulmonary window, 42, 54
  arteriovenous, 172
  atrial septal defect, 37, 42, 147, 212, 215, 216
  bidirectional, 35, 52
  double outlet right ventricle, 41
  Eisenmenger's syndrome, 53
  intra-cardiac, 1
  left-to-right, 36
  partial anomalous venous drainage, 37
  patent ductus arteriosus, 37, 44
  pseudo truncus arteriosos, 54
  quantitation, 4, 50, 51
  ratio-of-areas method, 52
  right-to-left, 33, 39, 47, 53
  scintigraphic study, 42, 44, 47
  serial assessment, 52
  single probe study, 36, 37
  tetralogy of Fallot, 53
  transposition, great vessels, 40
  truncus arteriosus, 44
  ventricular septal defect, 36, 42
 supra valvular aortic stenosis, 56
 triscuspid atresia, 54
Congestive failure, 98, 125ff, 215
Contamination, 35, 139, 218
 overlap, 195
 see also Background
Contaminating-count, 138
Contractility, 123ff, 136, 234
Contrast enhancement, 280
Convolution, 226
Cor pulmonale, 142
Coronary arteriography, 155, 193, 196
Coronary artery disease, 98, 155-201, 278-282
 atherosclerosis, 183
 collateral flow, 155
 electrically silent, 156
 ischemia, 155-201, 278-282
 myocardial lactate production, 156
 occluded segment, 155
 small coronary vessels, 155
Coronary microcirculation, 196
Count contamination, "cross talk," 218, 35, 139; see also Background
Count loss, 251
Count rate, 245
 pulse pileup, 244

Creative phosphokinase, 279
Cross-talk, 8, 35, 128, 130, 139, 218; see also Background
Crystal nonuniformity, 240
129Cs, 157, 170, 266, 278
137Cs, 262
Cursors, 233
Cyclotron, 254, 262

dA/dt, 124, 129, 218
Data loss, 235, 240
Data rate, 135
Data smoothing, Fourier transforms, 240
 Hadamard transforms, 240
Dead time, 9, 12, 235, 238, 240, 250
Derandomization, 235, 238
Diastole, 125ff, 209, 211, 214
Diastolic image, 90, 129
Digital computer, 176, 233; see also Computer
Dilution curves, 15, 19, 22, 28, 33, 35, 125, 228
 indicator, 4
 secondary curve components, 23
Diphosphonate, 279
Dose radiation, 6, 136, 256
 gonadal, 35
 reduced patient, 136
dV/dt, 124, 125, 136, 234

Ebstein's anomaly, 56
Echocardiography, 101, 117
Efficiency, 124
Efficiency constant, 215
Efficiency factor, 130
Ejection fraction, 1, 4, 5, 123, 125, 128, 130, 137, 138, 139, 142, 144, 214, 215, 220, 234
Electrocardiography, 90, 126, 156, 278
Electron capture, 255
End-diastole, 125, 132, 165
End-diastolic volume, 139, 143, 210ff
End-systole, 125, 132, 210ff
Endocardium, 195
Epicardium, 195
Error, 1, 5, 6, 23, 25, 35, 126, 133, 136
Exercise, 107, 278

External precordial counting, 136ff, 170; see also Radiocardiography

False positive, 37
Fatty acids, 267
Fick principle, 51, 173
Fine structure, 124
Fitting, 29, 134
Flagging, 25; see also Region of interest
Flip-flopping, 235
"Flood" exposure, 240
Flow rate, 28, 210; see also Cardiac output, Regional myocardial perfusion
FORTRAN IV, 243
Fourier analysis, 242
 transforms, 206, 240
Fractional elution, 256
Frame rate, 125, 135, 234, 235, 239
Frame time, 135, 214, 215
Frequency response, 125, 234
Function, emptying, 124, 215
 linear filling, 215
 ventricular volume, 213

GaGa citrate, 279
Gamma-camera, 32, 207, 214, 232, 255, 279
 counting rate, 245
 multicrystal, 249
 video instrumentation, 232
Gated image, 129ff, 166
Gating, 90, 126, 128, 132, 165
Generator, 256
Great vessels, aorta, 4
 pulmonary artery, 4

Hadamard transform, 240
Half-life, 255
Heterogeneity, perfusion, 179; see also Akinesis
High-frequency, 12, 23, 125, 204, 208
Histogram, 15, 24, 238, 240
Histogram mode, 236
Hydrogen gas, 171
Hydroxyapatite, 279
Hyperkinetic cardiovascular states, 107
Hypertension, left ventricular

hypertrophy, 90
Hyperthyroidism, 107
Hypokinesis, 93, 167

Indicator dilution, 33, 202
 delay line, 229
 kurtosis, 228
 mean transit time, 228
 mixing, 205
 quantitative analysis, 233
 rate constant, 188
 shunt quantitation, 34
 skewness, 228
 Stewart-Hamilton, 137
 variance, 228
 washout curves, 188
Indicator fractionation principal, 157
 extraction ratio, 157
 Sapirstein, 156
^{113m}In, 137
 transferrin, 260
Indium-DTPA, 261
Inert gas, 170, 172
 washout, 171
Infarction, subendocardial, 278
 transmural, 280; see also Myocardial infarction
Information theory, 234
Injection, EKG-triggered, powered, 5, 125
 technique, 5
Instrument, 231
 analog-to-digital coverter, 234
 data acquisition, 234
 digital computer, 233
 double-density tape recorder, 236
 light pen, 241
 magnetic disc, 236
 magnetic tape, 236
 matrix, 234
 persistence oscilloscope, 241
 phase mode, 236
 strip chart recorder, 233
 tape recorder, 236
Integration, 211
Intensive care, 152
Interface instrumental, 233
Interventricular septum, 5, Chapter 2
^{123}I, 261

^{124}I, 261
^{125}I-antipyrine, 173, 192
^{131}I, 136, 163, 165
Ischemia, 196
Isomeric transition, 255
Isometric mode, 241

^{42}K, 157ff
^{43}K, 157, 169, 266, 278
Kety-Schmidt formula, 172, 177, 188
^{81m}Kr, 266
^{85}Kr, 173, 190, 262, 263

Lagged normal density curve, 228
Laplace transforms, 206
Left anterior oblique, see Positioning
Left atrium, see Atrium left
Left-to-right shunt, quantitation, 50
 see also Congenital heart disease
Left ventricle, 5, 6, 8, 12, 25, 128, 230
Left ventricular failure, 98, 125ff., 149, 215
Left ventricular function, 123ff
Light pen, 238
Line-spread function, 157
Linear, 124
 response, 125
 scanner, 165
Linearity, 12
List mode, 234, 236, 237
Long-term storage, 235
Luetic aortitis, 96
Lung, 8, Chapter 3, 209

Maa, macroaggregate, 53, 268
 hazards; systemic embolism, 54
 see also Albumin
Magnetic disc, 236
Magnetic tape, 236
Matrix, 239
Mass balance equation, 205, 211, 212, 223
Mean transit time, 23, 137; see also Transit time
Medium frequency, 12, 208
Medium frequency response, 235
Microspheres, 163ff., 278
Minicomputer, 239, 242

Minimization program, 136
Mitral insufficiency, 98, 212, 215; see also Valvular heart disease
Mixing, 26, 208, 215
 convection, 206
 diffusion, 206
 dispersion, 206
 radiopaque, 208
Model, mathematical, 33, 202, 204, 211, 220
 compartment, 204, 205, 207, 208, 228
 constraints, 204
 delay, 205
 descriptive, 202, 228
 deterministic, 202, 228
 dispersion, 205, 228
 empirical, 228
 equations, 213
 four-parameter, 230
 lagged normal density, 206
 log-normal, 228
 mixer, 205
 multiple component, 204, 207
 parameter, 202, 211
 parameter estimation, 203
 parameter sensitivity, 204
 pulsatile, 208
 random walk, 228
 series mixer, 206
 single component, 204
 unique solution, 203
Molecular sieve, 263
Monoexponential, 196
 equation, 177
Multicrystal camera, 124, 195, 232, 249; see also Autofluoroscope
Multiple component, 202
Myocardial agents, 266, 278
Myocardial hypertrophy, 74
Myocardial imaging, 169, 278
Myocardial infarction, 184, 185, 278, 280
Myocardial oxygen consumption, 180
Myocardial perfusion, see Regional myocardial perfusion
Myocardial wall thickness, 5

^{13}N, 262, 265, 278
^{24}Na, 32
NA+4, 267

Narrow-angle scatter, 255
Nonlinearity, 240
Nonuniformity correction, 9, 240
Normal heart, 4
 anatomical features, 6
Nuclear reaction, low energy, 254
 medium energy, 254
Nutrient myocardial blood flow, 186, 189, 192; see also Regional myocardial perfusion

15O, 262, 265
Oscilloscope, refresh, 239
 storage, 239
Overlap, 175, see also Background

Papillary muscle, 67
Paradoxical myocardial kinetics, 90, 93
Parameter estimation, 203
Partition coefficient, 189, 192
Pericardial effusion, 117
Pertechnetate, see ^{99m}Tc
Polynomial function, 211
Polyphosphate, 279
Positioning, 5
 anterior, 5
 left anterior oblique (LAO), 5, 8, 96, 128, 157, 176
 left lateral, 5
 modified LAO, 8
 posterior oblique, 96
 right anterior oblique, 5, 67
Potassium, see K
Pulmonary artery, 6, 12, 25
 left, 8
Pulmonary blood volume (PBV), 1, 4, 28, 29, 137, 139, 142, 215
Pulmonary edema, 144
Pulmonary embolus, 103, 140, 142, 145, 146
Pulmonary hypertension, 103, 145, 146
Pyrophosphate (PYP), 280

Quantitative, 2
Quantilation of shunts, 4, 50, 53
 Fick, 51
 x/y ratio, 51

Radiation exposure, 6, 35, 136

Index

Radioactive gases, 256, 262;
   see also ^{133}Xe, ^{85}Kv
Radioactive particles, 163
 dual-label techniques, 164
 microspheres, 165
Radiocardiography, 143, 145,
   147, 152, 226
 atrial septal defect, Chapter
   5, 147
 cardiomyopathy, 144
 ejection fraction, 129, 144
 end diastolic volume, 143
 left ventricular failure, 149
 mitral insufficiency, 143
 pulmonary hypertension, 145
 pulmonary edema, 144
 pulmonary embolism, 145, 146
 pulmonary hypertension, 146
 see also Single probe study
Radiopharmaceuticals, 6, 25ff.
 dose, 6, 35, 136
Ratemeter, analog, 136
 digital, 137
^{86}Rb, 157, 266
Recirculation, 33, 36, 42, 137
Region of interest (ROI), 8,
   12, 15, 25, 125, 126, 129,
   136, 138, 218, 219, 240, 241,
   242
 flagging, 238
Regional Myocardial Perfusion
   (RMP), 163, 176, 184, 186,
   189, 192, 194, 196, 266
 abnormal wall motion, 167
 angina, 184
 aneurysm, 184
 atrial pacing, 184
 bypass procedures, 168
 capillary blood flow, 156, 172
 collateral blood flow, 186
 heterogeneity, 179
 hypertension, 181
 hypoperfusion, 167
 imaging, 168
 infarction, see Myocardial
   infarction
 inhalation, 170
 intracoronary injection, 161,
   170
 Kety-Schmidt, 170
 left ventricular hypertrophy,
   181
 myocardial hypertrophy, 181

 nutrient blood flow, 186, 189,
   192
 revascularization, 156
 subendocardial ischemia, 196
 tissue capillary blood flow,
   172
 washout curves, myocardial, 176
 washout techniques, 156
Regional ventricular dysfunction, 90, 93, 167, 179
Resolution, 234, 245, 247, 249
 spatial, 235, 239, 244
 temporal, 234, 235
Right atrium, see Atrium, right
Right-to-left shunt, see Congenital heart disease
Right ventricle, 14, 15
Rubidium, see Rb
Runge-Kutta integration, 214

Scintillation camera, 165, 196,
   232, 233, 235, 239, 244
 multiple-crystal, 12, 19, 66,
   174, 250
Screening, 3, 58, 119
Sedation, 5
Selective injection, 36
Septal hypertrophy, 74
Septum, interventricular, 8
Serial studies, 58, 132, 256
Simulation, 204, 205
Single probe (external precordial
   counting), 35, 39, 136, 137,
   150, 170, 215, 231
 anomalous pulmonary venous
   drainage, 37, 40
 atrial septal defect, 37
 coarctation of the aorta, 37
 method, 124
 pulmonary stenosis, 40
 transposition, 40
Sinus of Valsalva, 93
Slug injection, 226; see also
   Bolus
Smooth, 240
Sn (II), 256
^{113}Sn-In, tin-indium generator,
   260
Sodium, see Na
Software, 239, 244ff.
 routine functions, 239
Solubility coefficient, 172
Spatial resolution, 234

Spatial distortion, 240
Spatial volume methods, 132
Stewart-Hamilton, 226
Strip-chart recorder, 231
Stroke volume, 4, 123, 125, 139, 213
Subaortic stenosis, 77
 idiopathic, 77
Subendocardial infarction, 280
Superimposition, 23, 35; see also Overlap, Background
Superior vena cava, 6, 15
 obstruction of, 107
Systole, 210, 211, 214
Systolic image, 90, 132

^{201}T, 267
Tape recorder, 238
^{99m}Tc (Pertechnetate), 6, 35, 123, 128, 158, 163, 165, 170, 256
 albumin, human serum (TcHSA), 6, 26, 163
 erythrocytes, 257
^{99m}Tc-phosphate, 279
^{99m}Tc-pyrophosphate, 279
^{99m}Tc-tetracycline, 268, 279
Television camera, 232
Temporal resolution, 234
Temporal separation, 130, 256
Thallium-201, see ^{201}T
Time-activity curves, 15; see also Indicator dilution
Time constants, 27, 136
Time frame, 135, 214, 215, 234, 236, 239
Tin, see Sn
Transit time, 4, 125, 128, 204
 appearance, 22
 calculation of, 15
 mean transit, 19, 22
 peak time, 23
Tracer concentration, 210, 218
Tracer mass, 218, 226

Transmission source, 5
Transmural infarction, see Myocardial infarction
Tritiated water, 173

Unique solution, 203

Valvular heart disease, 6
 aortic insufficiency, 52, 83
 aortic stenosis, 74
 mitral commisurotomy, 67
 mitral insufficiency, 52, 67, 212
 mitral stenosis, 66
 mixed valvular lesions, 84
 pulmonic insufficiency, 84
 tricuspid insufficiency, 84
 vein bypass grafts, 163
Ventricle, 125
 left, 5, 6, 8, 12, 15, 25, 128, 230
 right, 5, 6, 8, 230
Ventricular diastolic volume, 1, 4, 139, 142, 143
Ventricular failure, 1, 74
Ventricular septal defect, quantitative study, 50
Videoscintiscope (VTV), 66, 232
Volume, end diastolic ventricular, 1, 4, 139, 142, 143
 end systolic, 4, 90, 125, 132
 pulmonary blood, 1, 4, 28, 29, 137, 139, 142, 215

Washout, 172, 173
 diffusion limited, 172
Washout curves, 170
Washout techniques, clearance, 156

^{123}Xe, 261
^{133}Xe, 170-201, 262, 264, 278

Zooming, 234